The Application of Science in Environmental Impact Assessment

T0199792

This book charts the history of the application of science in environmental impact assessment (EIA) and provides a conceptual and technical overview of scientific developments associated with EIA since its inception in the early 1970s.

The Application of Science in Environmental Impact Assessment begins by defining an appropriate role for science in EIA. From here it goes on to reflect more closely on empirical and deductive biophysical sciences as they relate to well-known stages of the generic EIA process and explores whether scientific theory and practice are at their vanguard in EIA and related applications. Throughout the book, the authors reflect on biophysical science as it applies to stages of the EIA process and also consider debates surrounding the role of science as it relates to political and administrative dimensions of EIA. Based on this review, the book concludes that improvements to the quality of science in EIA will rely on the adoption of stronger participatory and collaborative working arrangements.

Covering key topics, including foundational scientific guidance materials, frameworks for implementing science amid conflict and uncertainty, and emerging ecological concepts, this book will be of great interest to students, scholars, and practitioners of EIA.

Aaron J. MacKinnon has a Master of Environmental Studies from the School for Resource and Environmental Studies, Faculty of Management, Dalhousie University, Canada.

Peter N. Duinker is a professor at the School for Resource and Environmental Studies, Faculty of Management, Dalhousie University, Canada.

Tony R. Walker is an assistant professor at the School for Resource and Environmental Studies, Faculty of Management, Dalhousie University, Canada.

Routledge Focus on Environment and Sustainability

The Application of Science in Environmental Impact Assessment
Aaron MacKinnon, Peter Duinker and Tony Walker

The Application of Science in Environmental Impact Assessment

Aaron J. MacKinnon,
Peter N. Duinker and
Tony R. Walker

Routledge
Taylor & Francis Group

LONDON AND NEW YORK

First published 2018
by Routledge

2 Park Square, Milton Park, Abingdon, Oxfordshire OX14 4RN
52 Vanderbilt Avenue, New York, NY 10017

Routledge is an imprint of the Taylor & Francis Group, an informa business

First issued in paperback 2019

British Library Cataloguing-in-Publication Data
A catalogue record for this book is available from the British Library

Library of Congress Cataloging-in-Publication Data
A catalog record for this book has been requested.

ISBN: 978-0-8153-8729-9 (hbk)
ISBN: 978-0-367-34019-3 (pbk)

Typeset in Times New Roman
by Apex CoVantage, LLC

Contents

Abbreviations

AEAM	Adaptive environmental assessment and management
AVHRR	Advanced very high resolution radiometer
BACI	Before-after-control-impact
CCCEAC	Committee on Climate Change and Environmental Assessment in Canada
CEAA	Canadian Environmental Assessment Agency
CEQ	Council on Environmental Quality
EIA	Environmental impact assessment
EIS	Environmental impact statement
ESSA	Environmental and Social Systems Analysts Ltd
ERL	Environmental Resources Ltd
GIS	Geographic information system
GPS	Global Positioning System
IBP	International Biological Program
IGBP	International Geosphere Biosphere Program
LIDAR	Light imaging, detection, and ranging
LTER	Long Term Ecological Research Network
MEA	Millennium Ecosystem Assessment
MODIS	Moderate resolution imaging spectroradiometer
NAS	National Academy of Sciences
NASA	National Aeronautics and Space Administration
NGO	Non-governmental organization
NOAA	National Oceanic and Atmospheric Administration
NRC	National Research Council
NUSAP	Numerical, unit, spread, assessment, pedigree
PCB	Polychlorinated biphenyl
SFM	Sustainable forest management
TM	Thematic Mapper
UNCED	United Nations Conference on Environment and Development
UNFCCC	United Nations Framework Convention on Climate Change
VEC	Valued ecosystem component
VHF	Very high frequency
WCED	World Commission on Environment and Development
WCRP	World Climate Research Program

1 Introduction

1.1 Problem and context

Formal environmental impact assessment (EIA) processes were established by governments in North America during the early 1970s to satisfy growing public concerns over the environmental impacts of uncontrolled industrial and economic development. This marked the beginning of a vast new enterprise of environmental decision-making, complete with regulatory requirements, scientific contributions, and participatory processes. Today, EIA everywhere appears to be in a state of crisis, as scholars and practitioners alike continue to find the entire process seriously wanting (e.g., Ross et al. 2006). While some (e.g., Doelle 2012; Gibson 2012) criticize governments for ineffective administrative processes, others (e.g., Doelle and Sinclair 2006; Stewart and Sinclair 2007) find much to be improved upon in participatory mechanisms. Still others (e.g., Fairweather 1994; Treweek 1995; Warnken and Buckley 1998; Greig and Duinker 2011) find major fault in the scientific dimension of EIA. The situation observed today is puzzling, since the basic principles of competent EIA were published decades ago (e.g., Sadler 1996). Indeed, strong scientific principles for EIA were established in the early literature of the 1970s and 1980s (e.g., Holling 1978; Beanlands and Duinker 1983). Despite a growing body of literature outlining the basic requirements of competent EIA, it appears the EIA community continues to struggle with undertaking useful and defensible scientific investigations in EIA practice.

What was the original purpose of EIA? During the 1970s, the language used in EIA laws and policies was that of environmental protection, i.e., protection of important environmental values. To this end, administrative bodies were established by governments to translate new EIA laws and policies into regulatory procedures and requirements. Results of these early implementation efforts, however, were promptly challenged by members of the scientific community (e.g., Andrews 1973; Carpenter 1976; Schindler 1976) who observed that a founding principle of EIA – science in the service of environmental decision-making – was not finding application in EIA practice. Many scientists observed that existing procedural frameworks for EIA had favoured production of comprehensive but superficial environmental descriptions, rather than incisive environmental impact predictions. It was concluded, therefore, that EIA was not providing decision-makers with the critical knowledge of impacts needed to achieve EIA's original aim: protection of important environmental values.

Recognizing the need to operationalize EIA's scientific requirement, government agencies, universities, and non-governmental organizations (NGOs) in North America mobilized prominent research scientists to develop rigorous investigative frameworks for conducting EIAs (e.g., Holling 1978; Ward 1978; Munn 1979). This early guidance – collectively inspired by the interdisciplinary science of ecology – established foundational principles and protocols for predicting the biophysical impacts of development alternatives.

During the 1980s, researchers and practitioners continued to elaborate strong scientific principles and protocols for conducting EIAs (e.g., Duinker and Baskerville 1986; NRC 1986a; Walters 1986; Duinker 1989). Despite these ongoing conceptual and technical developments in the literature, the overall adequacy of EIA practice remained widely contested (e.g., Rosenberg et al. 1981; Clark et al. 1983; Hollick 1986; Culhane 1987; Fairweather 1989). In response to ongoing weaknesses in EIA practice, government agencies in North America funded major initiatives – one American (Caldwell et al. 1982) and one Canadian (Beanlands and Duinker 1983) – to investigate possibilities for improving the scientific basis for EIA. While both reports identified some important technical advancements in EIA practice since the early 1970s, they concluded that ongoing science challenges in EIA could be largely attributed to organizational and political factors. Both reports observed that existing review formats had fostered adversarial relations between those responsible for preparing EIAs (proponents and consultants) and those responsible for reviewing them (EIA administrators, legal professionals, research scientists, public interest groups, and other stakeholders). This had created an atmosphere of conflicting expectations, widespread confusion, and general frustration within the EIA community. Consequently, both reports observed the need to move away from adversarial review formats to focus on building more collaborative arrangements for EIA design and preparation.

In the Canadian project's final report, Beanlands and Duinker (1983) outlined a series of so-called requirements for conducting ecological studies in EIA. Most notably, they proposed the early identification of valued ecosystem components (VECs) in order to guide subsequent research activities. The VEC concept would go on to become a widely influential device for focusing assessments on stakeholder-relevant issues. To avoid the problem of costly and frustrating adversarial reviews, Beanlands and Duinker (1983) also proposed the establishment of a two-stage scientific review process. This process would rely on close collaboration between scientists, proponents, and consultants in developing mutually agreeable research designs prior to their implementation and final review. We consider the Beanlands and Duinker (1983) report, together with the literature of the 1970s, to provide the basic foundations for designing and implementing rigorous scientific investigations in EIA practice.

By the early 1990s, it was clear that the ultimate purpose of EIA had been broadly recast in terms of sustainable development (WCED 1987; Sadler 1996). In a major international review of EIA practice, Sadler (1996) observed the need to reconcile political, scientific, and administrative expectations within the EIA community to foster a more coherent and unified process for securing sustainable development decisions. In the project's final report, Sadler (1996) targeted four stages of the generic EIA process for immediate and cost-effective improvements: scoping, significance evaluation, document reviews, and follow-up monitoring. Consistent with Sadler's (1996) recommendations, researchers and practitioners of EIA have continued to elaborate strong approaches related to scoping (e.g., Mulvihill and Jacobs 1998; Mulvihill and Baker 2001; Mulvihill 2003), scenario-building (e.g., Duinker and Greig 2007), research collaboration (e.g., Greig and Duinker 2011), significance evaluation (e.g., Lawrence 2007a, 2007b, 2007c), effects monitoring (e.g., Arts 1998; Morrison-Saunders and Arts 2004), and biodiversity assessment (e.g., Nelson and Serafin 1991; Gontier et al. 2006).

Despite these ongoing developments in the literature, a considerable gap appears to persist between proposed ideals and their practical implementation. We maintain that if the global EIA enterprise is to fulfill its ultimate purpose – to contribute to a sustainable pattern of development by protecting VECs – researchers and practitioners around the world must strive to adopt a collaborative, participatory, and scientifically rigorous approach to conducting EIAs.

1.2 Objectives

This review aims first to provide a major summary and synthesis of scientific developments associated with EIA since the early 1970s, as evidenced in the peer-reviewed literature. In addition to key scientific developments, the review considers important political, organizational, and administrative factors related to implementation of science in EIA. The second objective is to judge, on the basis only of evidence in the peer-reviewed literature, whether scientific theory and practice are at their vanguard in EIA-related applications.

1.3 Approach, scope, and outline

The review focuses broadly on biophysical science as it applies to the EIA process. We expressly limit ourselves to biophysical domains of understanding, recognizing that social science is also important in much of EIA around the world but that applications of social science in EIA would require their own review of theory. Our use of the word 'science' from now on, without further specification, implies biophysical science.

Our conception of science is a broad one: it is a systematic way of acquiring and creating knowledge, and it also represents the body of knowledge so developed. Some call it a way of knowing, one that complements other ways of knowing such as lived experience and faith. Science is often viewed as anchored in the methods of formal observation and experimentation, but we would add a deductive component because science in the service of advising decision-makers about potential environmental outcomes of decision alternatives is a challenge of foresight. This latter element of science calls upon formal approaches to hypothesis creation, the testing of which frequently requires long-term, intensive measurement. Still others (e.g., Chapman 2016) argue that science without strong relationships between suppliers and demanders of knowledge will not generate usable insight.

It behoves us to explain how we see human values in the context of science. We do not subscribe to the view that science is a purely objective pursuit about an objectively determinable reality. Enough literature (e.g., Chapman 2016) has long dispelled that myth. Science is accomplished only by humans, and all humans have values. We prefer a conception of 'value' in this context as that which an individual or collection of individuals (i.e., an organization) considers important in relation to some aspect of life (Ordóñez et al. 2017). Early treatises on science in EIA (e.g., Beanlands and Duinker 1983) sought to distinguish between public values and scientists' values. That dichotomy, perhaps useful at the time, failed to acknowledge the ubiquity of diverse values characterizing the whole process of EIA – the legal framework, the process design, the application to disparate development decisions, and more. If the EIA enterprise is legitimately imagined as having interacting elements of administrative process, politics, and science, then the value sets of all the people engaged in these elements are relevant. Where applicable, we try to elucidate the value-laden nature of the scientific concepts we review in the book.

Determining what topics to include in our review and what order to use in laying them out has been a challenge. Part of the problem is that the conceptual domain of science in EIA can be tackled using a variety of lenses, each of which will expose problems in a different and potentially useful way. We felt it important first to examine what the process of EIA, through its entire evolution from 1970 to the present, expects from biophysical science. Since EIA is about more than just science, the other major elements being administrative process and political engagement, it seemed important to begin by examining how science in EIA can help to foster sustainable and mutually agreeable development decisions.

After situating science within the broader EIA enterprise, we decided to progress as follows: (a) from the highest-order concepts applicable to an examination of science in EIA to the most specific and (b) from the early years of EIA to the present. That called for an initial exploration of the scientific foundations of EIA. Those foundations arise from the 1970s literature that tried to guide scientific work in EIA. We then examine four recent conceptual frameworks for undertaking scientific investigations in the face of conflict and uncertainty: adaptive management, post-normal science, transdisciplinary imagination, and citizen science. From there we transition into eight emerging and contemporary ecological concepts for EIA. These are higher-order concepts that, in our view, have strong potential relevance to how science is practiced in the EIA context. These concepts strike us as particularly relevant to the EIA process and have emerged and gained prominence only after the EIA process became established as a mainstream decision-making process of governments worldwide.

Our final major substantive chapter examines the links between science and the major stages of the EIA process. We identify important scientific principles, protocols, and techniques as they apply to each stage of the process. The stages we chose are common to most EIA processes around the world: scoping, baseline characterization, impact prediction, significance evaluation, decision-making, and follow-up. The book concludes with a synthesis of major scientific developments associated with EIA since the early 1970s, highlighting important challenges and opportunities for practical implementation.

References

Andrews, R.N.L. 1973. A philosophy of environmental impact assessment. *Journal of Soil and Water Conservation* 28: 197–203.
Arts, J. 1998. *EIA Follow-up – On the Role of Ex Post Evaluation in Environmental Impact Assessment.* Groningen: Geo Press. 558 p.
Beanlands, G.E., and Duinker, P.N. 1983. *An Ecological Framework for Environmental Impact Assessment in Canada.* Halifax: Institute for Resource and Environmental Studies, Dalhousie University. 132 p.

Caldwell, L.K., Bartlett, R.V., Parker, D.E., and Keys, D.L. 1982. *A Study of Ways to Improve the Scientific Content and Methodology of Environmental Impact Analysis*. Bloomington: School of Public and Environmental Affairs, Indiana University. 453 p.

Carpenter, R.A. 1976. The scientific basis of NEPA – is it adequate? *Environmental Law Reporter* 6: 50014–50019.

Chapman, K. 2016. *Complexity and Creative Capacity: Rethinking Knowledge Transfer, Adaptive Management and Wicked Environmental Problems*. New York: Earthscan (Routledge). 244 p.

Clark, B.D., Bisset, R., and Tomlinson, P. 1983. *Post-Development Audits to Test the Effectiveness of Environmental Impact Prediction Methods and Techniques*. Aberdeen: PADC Environmental Impact Assessment and Planning Unit, University of Aberdeen. 47 p.

Culhane, P.J. 1987. The precision and accuracy of U.S. environmental impact statements. *Environmental Monitoring and Assessment* 8(3): 217–238.

Doelle, M. 2012. CEAA 2012: The end of federal EA as we know it? *Journal of Environmental Law and Practice* 24(1): 1–17.

Doelle, M., and Sinclair, A.J. 2006. Time for a new approach to public participation in EA: Promoting cooperation and consensus for sustainability. *Environmental Impact Assessment Review* 26(2): 185–205.

Duinker, P.N. 1989. Ecological effects monitoring in environmental impact assessment: What can it accomplish? *Journal of Environmental Management* 13(6): 797–805.

Duinker, P.N., and Baskerville, G.L. 1986. A systematic approach to forecasting in environmental impact assessment. *Journal of Environmental Management* 23: 271–290.

Duinker, P.N., and Greig, L.A. 2007. Scenario analysis in environmental impact assessment: Improving explorations of the future. *Environmental Impact Assessment Review* 27(3): 206–219.

Fairweather, P. 1989. Environmental impact assessment: Where is the science in EIA? *Search* 20(5): 141–144.

Fairweather, P. 1994. Improving the use of science in environmental assessments. *Australian Journal of Zoology* 29(3–4): 217–223.

Gibson, R.B. 2012. In full retreat: The Canadian government's new environmental assessment law undoes decades of progress. *Impact Assessment and Project Appraisal* 30(3): 179–188.

Gontier, M., Balfors, B., and Mörtberg, U. 2006. Biodiversity in environmental assessment – current practice and tools for prediction. *Environmental Impact Assessment Review* 26(3): 268–286.

Greig, L.A., and Duinker, P.N. 2011. A proposal for further strengthening science in environmental impact assessment in Canada. *Impact Assessment and Project Appraisal* 29(2): 159–165.

Hollick, M. 1986. Environmental impact assessment: An international evaluation. *Environmental Management* 10(2): 157–178.

Holling, C.S., editor. 1978. *Adaptive Environmental Assessment and Management.* Toronto: John Wiley & Sons. 377 p.

Lawrence, D.P. 2007a. Impact significance determination – designing an approach. *Environmental Impact Assessment Review* 27(8): 730–754.

Lawrence, D.P. 2007b. Impact significance determination – back to basics. *Environmental Impact Assessment Review* 27(8): 755–769.

Lawrence, D.P. 2007c. Impact significance determination – pushing the boundaries. *Environmental Impact Assessment Review* 27(8): 770–788.

Morrison-Saunders, A., and Arts, J. 2004. *Assessing Impact: Handbook of EIA and SEA Follow-Up.* London: Earthscan. 338 p.

Mulvihill, P.R. 2003. Expanding the scoping community. *Environmental Impact Assessment Review* 23(1): 39–49.

Mulvihill, P.R., and Baker, D.C. 2001. Ambitious and restrictive scoping: Case studies from Northern Canada. *Environmental Impact Assessment Review* 21(4): 363–384.

Mulvihill, P.R., and Jacobs, P. 1998. Using scoping as a design process. *Environmental Impact Assessment Review* 18(4): 351–369.

Munn, R.E., editor. 1979. *Environmental Impact Assessment: Principles and Procedures.* Second ed. Toronto: John Wiley & Sons. 190 p.

Nelson, J.G., and Serafin, R. 1991. *Biodiversity and Environmental Assessment.* Waterloo: International Union for the Conservation of Nature. 13 p.

NRC (National Research Council). 1986a. *Ecological Knowledge and Environmental Problem-Solving: Concepts and Case Studies.* Washington, DC: National Academy Press. 388 p.

Ordóñez, C., Beckley, T., Duinker, P.N., and Sinclair, A.J. 2017. Public values associated with urban forests: Synthesis of findings and lessons learned from emerging methods and cross-cultural case studies. *Urban Forestry & Urban Greening* 25: 74–84.

Rosenberg, D.M., Resh, V.H., Balling, S.S., Barnby, M.A., Collins, J.N., Durbin, D.V., Flynn, T.S., Hart, D.D., Lamberti, G.A., McElravy, E.P., et al. 1981. Recent trends in environmental impact assessment. *Canadian Journal of Fisheries and Aquatic Sciences* 38(5): 591–624.

Ross, W.A., Morrison-Saunders, A., Marshall, R., Sánchez, L.E., Weston, J., Au, E., Morgan, R.K., Fuggle, R., Sadler, B., Ross, W.A., et al. 2006. Improving quality. *Impact Assessment and Project Appraisal* 24(1): 3–22.

Sadler, B. 1996. *Environmental Assessment in a Changing World: Evaluating Practice to Improve Performance.* Final Report of the International Study of the Effectiveness of Environmental Assessment. Hull: Canadian Environmental Assessment Agency. 248 p.

Schindler, D.W. 1976. The impact statement boondoggle. *Science* 192(4239): 509.

Stewart, J.M.P., and Sinclair, J.A. 2007. Meaningful public participation in environmental assessment: Perspectives from Canadian participants, proponents, and government. *Journal of Environmental Assessment Policy and Management* 9(2): 161–183.

Treweek, J. 1995. Ecological impact assessment. *Impact Assessment* 13(3): 289–315.

Walters, C.J. 1986. *Adaptive Management of Renewable Resources*. New York: Palgrave Macmillan. 374 p.

Ward, D.V. 1978. *Biological Environmental Impact Studies: Theory and Methods*. New York: Academic Press. 157 p.

Warnken, J., and Buckley, R. 1998. Scientific quality of tourism environmental impact assessment. *Journal of Applied Ecology* 35(1): 1–8.

WCED (World Commission on Environment and Development). 1987. *Our Common Future*. Oxford: Oxford University Press. 383 p.

2 Methods

In today's parlance, we conducted a comprehensive or narrative overview of the scholarly literature surrounding EIA-related science. As such, the review is both descriptive, in summarizing the contents of the items retrieved, and critical, in judging the quality or validity of the findings expressed therein. We began by building a hierarchical set or organizing themes and concepts to structure the review. Based on several decades of experience as practitioners and scholars of EIA, we were able to identify and group several key concepts that have shaped collective thinking about EIA-related science since the 1970s. At the finest level of detail, we identified particular kinds of tools and techniques (e.g., remote sensing, simulation modelling) that might be used at different stages of the EIA process (e.g., baseline characterization, impact prediction). At a much broader level, we identified several conceptual frameworks (e.g., adaptive management, post-normal science) that have shaped contemporary thinking about the implementation of science in EIA. We also identified a number of ecological concepts (e.g., biodiversity, climate change) that have begun to find their way into EIA-related scientific applications over the last few decades. Lastly, we decided that a review of the EIA-related scientific literature would not be complete without consideration for EIA's other two dimensions (i.e., political, administrative), so we included them as minor themes in the review as well.

We began obtaining relevant literature by conducting searches in three major electronic databases: Google Scholar, Web of Science, and Scopus. Database queries typically comprised specific scientific terms (e.g., indicator, simulation model) combined with generic EIA terms (e.g., environmental impact assessment). Such phrases were carefully constructed to query all three databases exhaustively, thereby ensuring comprehensive coverage of the literature. To ensure even coverage through time, we consistently sorted all search results by decade (e.g., 1970s, 1980s). In addition to searching electronic databases, we conducted hand searches of EIA textbooks, government reports, as well as official and semi-official guidance materials. As a final means of ensuring complete coverage, we consulted the literature-cited lists of all the authoritative articles and books retrieved through earlier searches. This last step in the retrieval process yielded a considerable volume of relevant material.

In our attempt to capture a wide range of literature, we deliberately adopted a broad definition of what constitutes EIA-related science. Because science in EIA spans many academic disciplines, we were forced to draw on literature from diverse scientific fields situated outside EIA, particularly the science of ecology. Where the literature offered important political and administrative perspectives on implementing science in EIA, we made sure to include them in the review as well.

Literature items obtained through searches (~5,000) were first organized into a hierarchical set of file folders based on important concepts and search terms (e.g., biodiversity, scoping, prediction). Each item was then carefully examined to determine its relevance to the review. As part of the vetting process, we constructed a spreadsheet database arraying relevant literature items (~700) against their contents based on keywords (e.g., climate change, resilience, prediction). In the end, our comprehensive approach to canvassing combined with our highly selective approach to screening ensured that far more literature was initially retrieved than was included in the final review.

Though we are aware that important advancements in EIA-related science may also be found in the regulatory filings associated with formal EIA practice, such grey literature was deemed outside the scope of this review. We do note, however, that many of the scholarly articles reviewed here did draw heavily upon such grey literature. It is our hope, therefore, that certain elements of EIA practice have made their way, if only indirectly, into this review. We also note that while the review draws heavily upon literature addressing the North American (especially American and Canadian) EIA experience, generalizations were made wherever possible in order to make overall lessons more applicable to the worldwide EIA enterprise.

3 Conceiving a role for science in EIA

3.1 Science and politics

Formal EIA processes around the world require various combinations of scientific investigation and public participation to support the ultimate goals of environmental protection and sustainable development. Because there is often little guidance on how to integrate scientific and political elements in practice, a divergence of expectations has emerged within the EIA community (Sadler 1996). In a recent survey of EIA practitioners, Morrison-Saunders and Sadler (2010) found that while most respondents recognized the fundamental need for both scientific and political inputs to EIA decision-making – especially with respect to achieving sustainable development decisions – they could not necessarily explain how science and politics should be integrated in practice.

There are multiple ways to place science and politics into a theoretical construction of EIA (e.g., Bartlett and Kurian 1999; Cashmore 2004). According to Cashmore (2004), the role of science in EIA can be seen along a spectrum ranging from EIA as applied science to EIA as civic science, with five distinct models identified within the two paradigms. In the applied science models (analytical science and environmental design), the methods of the natural sciences prevail. In the civic-science models (information provision, participation, and environmental governance), the emphasis is on stakeholder participation and value judgements. According to Cashmore (2004), the perceived role of science in EIA depends largely on one's views regarding the immediate purpose of EIA, i.e., how it interacts with decision-making processes (informing, influencing, or integrating). Cashmore (2004) argues that because science in EIA has merely sought to interact with decision-makers through passive information provision, it has largely failed to shape the final outcomes of their decisions. Cashmore (2004) suggests that demands for stronger participatory politics in EIA – aimed at actively influencing development decisions – have arisen from a growing awareness of science's failure to sway decision-makers. In short, Cashmore (2004) places science and politics at odds with one another in the battle to influence sustainable development decisions.

We prefer to understand the relationship between science and politics in EIA within the context of Lee's (1993) framework for pursuing sustainable development. In his landmark 1993 book entitled *Compass and Gyroscope: Integrating Science and Politics for the Environment*, Lee (1993) compellingly argues that a sustainable pattern of development is most effectively pursued through the integration of experimental science and participatory politics. To this end, Lee (1993) outlines a process of collective social learning that relies on adaptive management (Holling 1978; Walters 1986) to reduce scientific uncertainty, and principled negotiation to bound stakeholder conflict. For Lee (1993), adaptive management and principled negotiation together serve as complementary navigational aids (i.e., 'compass and gyroscope') in the pursuit of sustainable development. In the context of Lee's (1993) framework, EIA is conceived as an ongoing process of structured social learning in which experimental science, oriented by bounded stakeholder conflict, can provide useful and defensible knowledge to stakeholders and decision-makers in the collective pursuit of sustainable development. In other words, the immediate purpose of EIA – understood here as the creation and mobilization of reliable knowledge in support of sustainable development – is most effectively achieved through the fusion of adaptive management and principled negotiation.

3.2 Science inside and science outside

Around the world, independent reviews of EIA findings are typically required in cases of controversy and conflict. In such cases, independent review bodies may call upon research scientists to provide expert testimony during adversarial review hearings. While this sort of arrangement does allow research scientists to participate in the evaluation of EIA findings, it does not encourage them to collaborate with consultants and proponents during the design and implementation of scientific studies.

Greig and Duinker (2011) prefer to place research scientists into more constructive roles with respect to EIA practice. According to Greig and Duinker (2011), science outside EIA is needed to create robust ecological effects knowledge for practitioners, while science inside EIA is needed to apply that knowledge to generate reliable impact predictions for decision-makers. Greig and Duinker (2011) observe that because professional consultants working inside EIA are often faced with severe time and resource constraints, research scientists outside EIA are needed to assist in creating reliable ecological effects knowledge over longer time frames. Indeed, research scientists operating outside EIA have the capacity to mount long-term process studies in large ecosystems, undertake experimental cause-effect research for specific development types and VECs, and assemble general effects knowledge into models for predicting the impacts of development alternatives. To reciprocate, science inside EIA is able to provide practical applications and specific test cases for the refinement of general effects knowledge generated by science outside EIA.

To enable the sorts of organizational arrangements needed to support collaborative impact research, Greig and Duinker (2011) propose the establishment of multi-stakeholder research networks resembling those that have supported similar sustainable development initiatives (e.g., SFM Network 2016). Research networks intended for EIA would comprise unique partnerships between universities, governments, businesses, and NGOs according to specific development types, VECs, and regional contexts. These coordinated research networks would engage EIA practitioners in designing the targeted research programs needed to generate robust ecological effects knowledge, thereby contributing to sustainable development while reducing scientific uncertainty over time.

The position taken by Greig and Duinker (2011) – that the EIA community continues to struggle with scientific practice – stands in contrast with the findings of Morrison-Saunders and Bailey (2003) and Morrison-Saunders and Sadler (2010), who surveyed EIA practitioners on two occasions and found that they were in large measure pleased with the quality of science in EIA, but displeased with the importance placed on science at various stages of the EIA process. Greig and Duinker (2011) responded by observing that such findings are not surprising, since the science practiced inside EIA is under the direct purview of those very practitioners. Greig and Duinker (2011) also pointed out that EIA practitioners would obviously defend the quality of their own work and would naturally express disappointment when others did not place much importance on it. In short, Greig and Duinker (2011) maintain that substantial improvements can and should be made to the quality of science practiced inside EIA. To do so, however, will require the establishment of collaborative research networks linking scientific practice inside EIA with scientific research conducted outside EIA.

Barriers to achieving such collaborative partnerships are well known (e.g., Briggs 2006; Rogers 2006; Roux et al. 2006; Gibbons et al. 2008). According to Ryder et al. (2010), such obstacles arise from the fundamentally different perceptions that scientists, managers, and stakeholders have of science and scientific knowledge. To address this, several environmental laws and regulations have sought to standardize scientific contributions to resource and environmental decision-making by requiring that such decisions be based on so-called 'best available science' (e.g., Bisbal 2002; Glicksman 2008; Mills et al. 2009; Green and Garmestani 2012; Murphy and Weiland 2016). Oddly, such mandates have failed to include an explicit definition of what constitutes best available science or how it might be applied to environmental decision-making.

Attempts to elaborate on the concept (e.g., Bisbal 2002; Sullivan et al. 2006; Ryder et al. 2010; Green and Garmestani 2012) have observed that non-scientists tend to view science as a collection of information, whereas scientists themselves tend to see science as a systematic process of gathering and organizing knowledge into theories and testable hypotheses. Such ambiguity has led to disagreement over the precise meaning of the term 'best available science'. Proposing a way forward, Ryder et al. (2010) identified several core attributes for best available science. Ryder et al. (2010) also observed that these principles are clearly embedded in the science of adaptive management. In the words of Ryder et al. (2010): "The creation of interdisciplinary teams in an adaptive management framework is an essential process to identify research questions, create a 'taxonomy' of available information and facilitate the incorporation of new scientific information as it becomes available". It is clear, then, that the creation of the best scientific information inside EIA will rely not only on the application of the best scientific tools and techniques available, but on the adoption of collaborative working arrangements that facilitate productive researcher-practitioner relationships.

3.3 Science, politics, and administration

Sinclair et al. (2017) write that all environment and resource decision-making can be broadly conceptualized in terms of three fundamental dimensions: (i) the administrative/regulatory dimension, which is dominated by government responsibilities, timelines, and required procedures to obtain development approval; (ii) the participatory/political dimension, which is dominated by stakeholder relations, conflict, power, and civic engagement; and (iii) the scientific/technical dimension, which is dominated by scientific protocols for developing and testing impact predictions. The scientific dimension (including both traditional and local ecological knowledge) aims to deliver the critical understanding of potential impacts that informs discussions and debates in the other two dimensions. In the context of Sinclair et al.'s (2017) framework, all three dimensions require satisfactory implementation if the central purpose of EIA – now commonly agreed upon as sustainable development – is to be achieved. Avoidance of the administrative/regulatory dimension may mean that strong results from scientific and participatory processes have no influence on government regulators. Poor performance in the participatory/political dimension may mean that robust scientific results within a strong regulatory framework go unheeded as communities rise up and oppose development. Weak science, even in situations of strong participatory processes and successful navigation of bureaucratic administration, may lead to developments that have undesirable environmental impacts.

In sum, science – conceived here as the production of reliable knowledge about the potential biophysical impacts of development alternatives – is an absolutely necessary yet wholly insufficient element of competent EIA. Indeed, formal EIA processes around the world, as defined by the regulatory requirements under which most EIA is conducted, typically demand a scientific contribution in the form of baseline characterizations, cause-effect research, impact predictions, and monitoring. Clearly, EIA is missing a key element if it is strictly a participatory and regulatory process. Even worse is when it is conceived as a strictly scientific process, for while influence without insight may be dangerous, insight without influence is hardly useful.

References

Bartlett, R.V., and Kurian, P.A. 1999. The theory of environmental impact assessment: Implicit models of policy making. *Policy and Politics* 27(4): 415–433.

Bisbal, G.A. 2002. The best available science for the management of anadromous salmonids in the Columbia River Basin. *Canadian Journal of Fisheries and Aquatic Sciences* 59: 1952–1959.

Briggs, S.V. 2006. Integrating policy and science in natural resource management: Why so difficult? *Ecological Management & Restoration* 7: 37–39.

Cashmore, M. 2004. The role of science in environmental impact assessment: Process and procedure versus purpose in the development of theory. *Environmental Impact Assessment Review* 24(4): 403–426.

Gibbons, P., Zammit, C., Youngentob, K., Possingham, H.P., Lindenmayer, D.B., Bekessy, S., Burgman, M., Colyvan, M., Considine, M., Felton, A., Hobbs, R.J., Hurley, K., McAlpine, C., McCarthy, M.A., Moore, J., Robinson, D., Salt, D., and Wintle, B. 2008. Some practical suggestions for improving engagement between researchers and policy-makers in natural resource management. *Ecological Management & Restoration* 9(3): 182–186.

Glicksman, R.L. 2008. Bridging data gaps through modeling and evaluation of surrogates: Use of the best available science to protect biodiversity under the National Forest Management Act. *Indiana Law Journal* 83: 465–527.

Green, O.O., and Garmestani, A.S. 2012. Adaptive management to protect biodiversity: Best available science and the Endangered Species Act. *Diversity* 4: 164–178.

Greig, L.A., and Duinker, P.N. 2011. A proposal for further strengthening science in environmental impact assessment in Canada. *Impact Assessment and Project Appraisal* 29(2): 159–165.

Holling, C.S., editor. 1978. *Adaptive Environmental Assessment and Management.* Toronto: John Wiley & Sons. 377 p.

Lee, K.N. 1993. *Compass and Gyroscope: Integrating Science and Politics for the Environment.* Washington, DC: Island Press. 243 p.

Mills, A., Francis, T., Shandas, V., Whittaker, K., and Graybill, J.K. 2009. Using best available science to protect critical areas in Washington state: Challenges and barriers to planners. *Urban Ecosystem* 12: 157–175.

Morrison-Saunders A., and Bailey, J. 2003. Practitioner perspectives on the role of science in environmental impact assessment. *Environmental Management* 31(6): 683–695.

Morrison-Saunders, A., and Sadler, B. 2010. The art and science of impact assessment: Results of a survey of IAIA members. *Impact Assessment and Project Appraisal* 28(1): 77–82.

Murphy, D.D., and Weiland, P.S. 2016. Guidance on the use of best available science under the US Endangered Species Act. *Environmental Management* 58: 1–14.

Rogers, K.H. 2006. The real river management challenge: Integrating scientists, stakeholders and service agents. *River Research and Application* 22: 269–280.

Roux, D.J., Rogers, K.H., Biggs, H.C., Ashton, P.J., and Sergeant, A. 2006. Bridging the science-management divide: Moving from unidirectional knowledge transfer to knowledge interfacing and sharing. *Ecology and Society* 11(1): 4.

Ryder, D.S., Tomlinson, M., Gawne, B., and Likens, G.E. 2010. Defining and using 'best available science': A policy conundrum for the management of aquatic ecosystems. *Marine and Freshwater Research* 61: 821–828.

Sadler, B. 1996. *Environmental Assessment in a Changing World: Evaluating Practice to Improve Performance.* Final Report of the International Study of the Effectiveness of Environmental Assessment. Hull: Canadian Environmental Assessment Agency. 248 p.

SFM (Sustainable Forest Management) Network. 2016. *Sustainable Forest Management Network Legacy* [Internet]; [cited 2016 December 11]. Available from: http://sfmn.ualberta.ca/

Sinclair, A.J., Doelle, M., and Duinker, P.N. 2017. Looking up, down, and sideways: Reconceiving cumulative effects assessment as a mindset. *Environmental Impact Assessment Review* 62: 183–194.

Sullivan, P.J., Acheson, J.M., Angermeier, P.L., Faast, T., Flemma, J., Jones, C.M., Knudsen, E.E., Minello, T.J., Secor, D.H., Wunderlich, R., and Zanetell, B.A. 2006. Defining and implementing best available science for fisheries and environmental science, policy, and management. *Fisheries* 31(9): 460–465.

Walters, C.J. 1986. *Adaptive Management of Renewable Resources.* New York: Palgrave Macmillan. 374 p.

4 Foundations of science in EIA

4.1 Early EIA methods

Early methods for conducting EIA have been comprehensively reviewed elsewhere in the literature (e.g., Bisset 1978; Canter 1983; Shopley and Fuggle 1984; Wathern 1984). These methods are generally divided into four broad categories:

  (i) overlays;
  (ii) matrices;
  (iii) checklists;
  (iv) networks.

The use of spatial overlays in EIA was first suggested by landscape architects at the University of Pennsylvania (McHarg 1969). Using this technique, analysts would visualize the spatial implications of alternative transportation routes by overlaying a series of shaded or coloured transparencies depicting important environmental factors with transparencies depicting development characteristics. In the 1970s, following the expansion of geographic information systems (GIS), rasterized data layers could be used in place of transparencies to quantitatively depict environmental factors and development characteristics. Using this approach, analysts assign numerical values to each raster-cell in a given data layer, overlay multiple data layers, and then aggregate the overlapping numbers to obtain spatially explicit estimates of impact magnitude/significance.

The most well-known of the early EIA methods is undoubtedly the Leopold matrix, developed and published by the US Geological Survey (Leopold et al. 1971). This large rectangular array lists 100 generic development actions on the horizontal axis against 88 generic environmental components on the vertical axis, for a total of 8,800 potential interactions. By mentally 'overlaying' project proposals onto site descriptions, analysts assign separate magnitude and significance ratings (1–10) to all potentially relevant interactions. These dimensionless ratings may then be summed to obtain a total impact score.

Weighted-checklist methods for EIA include the Environmental Evaluation System, developed by Battelle Laboratories for the US Bureau of Land Reclamation (Dee et al. 1973). This method employs expert-derived value functions and weights to transform 78 estimated water quality parameters into 18 dimensionless water quality scores, each representing the condition of a generic environmental component. In this framework, an impact is defined as the difference between estimates of current (unaffected) and future (affected) environmental conditions.

The use of causal network diagrams in EIA was first suggested by a student at the University of California (Sorensen 1971). Using so-called Sorensen networks, analysts draw linkages between development actions and linear sequences of environmental components (typically two or three of these). They then qualitatively describe expected development-induced changes in those components. The original Sorensen networks were intended to depict relationships linking generic development actions to qualitative changes in six categories of generic environmental components.

4.2 Challenges from the scientific community

Objections to early methods – particularly matrices and checklists – can be found throughout the early EIA literature (e.g., Andrews 1973; Lapping 1975; Bisset 1978; Holling 1978; Munn 1979). From a scientific perspective, it was observed that matrices and checklists do not provide rigorous investigative protocols for predicting the biophysical impacts of development alternatives. This observation was associated with the fact that matrices and checklists cannot depict spatial and temporal dimensions of the environment, or functional relationships among environmental entities. Instead, matrix and checklist approaches were seen to favour the production of shallow and descriptive EIAs, a phenomenon attributed to the lists of generic environmental components used to guide field surveys and data collection. The compartmentalization of environmental entities into discrete and unrelated categories was seen to discourage interdisciplinary collaboration among specialists, further contributing to the production of fragmented and descriptive EIAs. It was concluded that the resulting impact scores are based largely on implicit expert opinion, rather than observable properties of the biophysical environment, and therefore do not represent valid or testable measures of environmental impact.

From both scientific and administrative perspectives, Andrews (1973) argued that comprehensive EIA methods such as the Leopold matrix were the product of unrealistic expectations held by administrators for complete environmental information. He argued that early regulatory approaches to implementing EIA were often based on the assumption that more information alone leads to better decisions, a notion widely reinforced by the threat of judicial challenge. Andrews (1973) further argued that the Leopold matrix – being unable to generate explicit impact predictions using scientific protocols – had failed to provide decision-makers with the critical insights needed to choose carefully among development alternatives. He concluded, therefore, that such early methods were unable to fulfill EIA's intended role in fostering development decisions that protect important environmental values.

Early methods for conducting EIAs were deemed problematic from a political perspective as well. Bisset (1978) argued that by failing to decouple the quantification of impact significance (based on implicit expert opinion) from the estimation of impact magnitude (based on perceptible attributes of the environment), early matrix and checklist methods impinged on the interpretation and evaluation of assessment results by other EIA participants. In other words, by concealing the value-based judgements of those who prepare assessments in the form of numerical impact scores, early matrix and checklist methods ultimately served to constrain wider stakeholder influence over development decisions.

4.3 First generation scientific guidance

The first generation of scientific guidance materials for EIA emerged in the late 1970s (e.g., Holling 1978; Ward 1978; Munn 1979). This guidance collectively advocated an approach to EIA based on collaborative problem-structuring, simulation modelling, quantitative measurement, and experimental manipulation. The most influential of these early frameworks – outlined by Holling (1978) and colleagues at the University of British Columbia – is composed of a set of concepts, protocols, and techniques collectively referred to as adaptive environmental assessment and management (AEAM). To overcome reactive, fragmented, and descriptive approaches to EIA, Holling's (1978) framework sought to coordinate ongoing communication and collaboration among scientists and EIA practitioners beginning in the early stages of development design. To this end, the AEAM approach employs workshops and computer simulation modelling exercises to explicitly define problems surrounding proposed development alternatives. During early workshops, participants reach agreement on the most important components and relationships characterizing the ecosystems under study. Participants must also select perceptible attributes of key ecosystem components – referred to as performance indicators – to serve as measures of environmental change. As workshops progress, implicit understandings of cause-effect relationships are gradually quantified and translated into dynamic computer simulation models. These models are to provide a shared conceptual and technical framework for cause-effect research, impact prediction, and ongoing effects monitoring.

The immediate objectives of the AEAM approach are twofold: (i) to formulate defensible impact predictions that are useful to decision-makers, and (ii) to explicitly acknowledge and reduce scientific uncertainty over time through continuous effects monitoring and model refinement. To effectively bridge functional gaps between development design, impact assessment, and regulatory approvals processes, AEAM workshops require close and ongoing collaboration between scientists, proponents, and regulators. In this way, EIA is understood as an ongoing, collaborative, and focused approach to development design and environmental decision-making, rather than a reactive, fragmented, and descriptive approach to evaluating a single preferred development design.

4.4 Provisions for scoping

In the late 1970s, attempting to overcome the proliferation of unfocused and overly descriptive EIAs, the US Council on Environmental Quality (CEQ) outlined the first formal requirements for scoping – an early and open process for determining the specific issues to be addressed in an EIA (CEQ 1980). Despite the provision of scoping requirements and the ongoing elaboration of scientific protocols for EIA during the 1980s (e.g., Duinker and Baskerville 1986; NRC 1986a; Walters 1986; Duinker 1989), scholars and practitioners continued to observe considerable weaknesses in the application of science in EIA practice (e.g., Rosenberg et al. 1981; Clark et al. 1983; Culhane 1987; Fairweather 1989). It was observed that mainstream EIA practice had continued to pursue comprehensive but superficial coverage of environmental entities, regardless of their relevance to development decision-making. It was also observed that impact predictions, if made at all, were generally comprised of vague statements about the likelihood of certain environmental conditions prevailing, with little or no accompanying discussion of uncertainty.

To explain the growing state of confusion and frustration surrounding the scientific adequacy of EIA practice, Carpenter (1983) pointed to a major divergence of expectations and capabilities within the EIA community. From both scientific and administrative perspectives, Carpenter (1983) observed that those responsible for reviewing EIAs (administrators, legal professionals, public interest groups, and other stakeholders) had developed unrealistic expectations of what could reasonably be accomplished by those responsible for preparing EIAs (proponents and consultants). He also observed that the wider EIA community had thus far failed to acknowledge (and agree on coordinated measures to reduce) the profound uncertainty surrounding complex ecosystems. Instead, he noted that most EIA administrators, legal professionals, and laypeople had come to expect complete and verified impact predictions from a fledgling environmental science operating within severe time and resource constraints.

Most importantly perhaps, Carpenter (1983) observed that the role of research scientists – i.e., those with the best understanding of ecological science and its practical limitations – was typically restricted to providing expert testimony during adversarial review hearings. The result, concluded Carpenter (1983), was a frustrating state of affairs in which research scientists were being asked to challenge or defend the quality of work prepared by proponents and their consultants, all within a judicial or quasi-judicial review framework that itself emphasized the legal and procedural requirements of EIAs.

To reconcile conflicting expectations within the EIA community, and to share the burden of reducing uncertainty surrounding complex ecosystems, Carpenter (1983) called for: (i) the design and implementation of long-term ecosystem monitoring programs, undertaken collaboratively by governments, proponents, and research scientists, and (ii) the establishment of formal scientific advisory and review committees, intended to assist in the development and review of targeted research programs for particular EIAs. In sum, Carpenter (1983) called for early and ongoing collaboration among all EIA participants, including members of the scientific community.

4.5 Second generation scientific guidance

In the 1980s, a second generation of scientific guidance materials – some North American (e.g., Beanlands and Duinker 1983; Duinker and Baskerville 1986; Duinker 1989) and some European (e.g., ERL 1981; ERL 1984; ERL 1985) – provided further insight into the application of science in EIA. In a major Canadian report on the subject, Beanlands and Duinker (1983) outlined a series of six so-called requirements for conducting ecological studies in the context of formal EIA:

(i) Identify the VECs on which the analysis will focus
(ii) Define a context for impact significance
(iii) Establish boundaries for analysis
(iv) Develop and implement a study strategy
(v) Specify the nature of predictions
(vi) Undertake monitoring

All six 'requirements' would pertain to the planning and design stages of the EIA process because, according to Beanlands and Duinker (1983), it is during these early stages that scientific improvements are most effectively realized. In summary, all EIAs should be required to: (i) "identify at the beginning of the assessment an initial set of valued ecosystem components to provide a focus for subsequent activities"; (ii) "define a context within which the significance of changes in the valued ecosystem components can be determined"; (iii) "show clear temporal and spatial contexts for the study and analysis of expected changes in valued ecosystem components"; (iv) "develop an explicit strategy for investigating the interactions between a project and each valued ecosystem component"; (v) "state impact predictions explicitly and accompany them with the basis upon which they were made"; and (vi) "demonstrate and detail a commitment to a well defined programme for monitoring project effects".

Crucial to focusing all subsequent stages of the research program would be the identification of VECs through scoping. The early identification of VECs was intended to provide an opportunity for all EIA participants to build consensus around the most important environmental values at stake. Practitioners would thereby be encouraged to design and implement targeted research programs aimed at providing useful and defensible insights into stakeholder-relevant issues. By encouraging collaboration early on, EIA preparers and reviewers would have an opportunity to reach agreement on the most appropriate research design prior to the commitment of time and resources, thereby avoiding costly and frustrating reviews upon completion of the assessment.

Since then, the Beanlands and Duinker (1983, 1984) proposals have been widely cited in both scholarly and practitioner literatures alike. While it is clear that the VEC concept has become an important device for focusing scientific investigations in EIA practice, uptake of the other five requirements by the practitioner community appears to have been sparse. In the remainder of this book, we build on the scientific foundations of the 1970s and 1980s by comprehensively reviewing and synthesizing insights from the relevant peer-reviewed literature.

References

Andrews, R.N.L. 1973. A philosophy of environmental impact assessment. *Journal of Soil and Water Conservation* 28: 197–203.

Beanlands, G.E., and Duinker, P.N. 1983. *An Ecological Framework for Environmental Impact Assessment in Canada*. Halifax: Institute for Resource and Environmental Studies, Dalhousie University. 132 p.

Beanlands, G.E., and Duinker, P.N. 1984. An ecological framework for environmental impact assessment. *Journal of Environmental Management* 18: 267–277.

Bisset, R. 1978. Quantification, decision-making and environmental impact assessment in the United Kingdom. *Journal of Environmental Management* 7: 43–58.

Canter, L.W. 1983. Methods for environmental impact assessment: Theory and application (emphasis on weighting-scaling checklists and networks). In: *Environmental Impact Assessment*. PADC Environmental Impact Assessment and Planning Unit, editor. The Hague: Nijhoff. 165–233 p.

Carpenter, R.A. 1983. Ecology in court, and other disappointments of environmental science and environmental law. *Natural Resources Law* 15: 573–596.

CEQ (Council on Environmental Quality). 1980. *Environmental Quality – The Eleventh Annual Report of the Council on Environmental Quality*. Washington, DC: Council on Environmental Quality. 497 p.

Clark, B.D., Bisset, R., and Tomlinson, P.1 983. *Post-Development audits to Test the Effectiveness of Environmental Impact Prediction Methods and Techniques*. Aberdeen: PADC Environmental Impact Assessment and Planning Unit, University of Aberdeen. 47 p.

Culhane, P.J. 1987. The precision and accuracy of U.S. environmental impact statements. *Environmental Monitoring and Assessment* 8(3): 217–238.

Dee, N., Baker, J., Drobny, N., Duke, K., Whitman, I., and Fahringer, D. 1973. An environmental evaluation system for water resource planning. *Water Resource Research* 9(3): 523–535.

Duinker, P.N. 1989. Ecological effects monitoring in environmental impact assessment: What can it accomplish? *Environmental Management* 13(6): 797–805.

Duinker, P.N., and Baskerville, G.L. 1986. A systematic approach to forecasting in environmental impact assessment. *Journal of Environmental Management* 23: 271–290.

ERL (Environmental Resources Ltd). 1981. *Methodologies, Scoping and Guidelines*. The Hague: Ministry of Public Housing, Physical Planning and Environmental Affairs. 103 p.

ERL (Environmental Resources Ltd). 1984. *Prediction in Environmental Impact Assessment*. The Hague: Ministry of Public Housing, Physical Planning and Environmental Affairs. 318 p.

ERL (Environmental Resources Ltd). 1985. *Handling Uncertainty in Environmental Impact Assessment*. The Hague: Ministry of Public Housing, Physical Planning and Environmental Affairs. 285 p.

Fairweather, P. 1989. Environmental impact assessment: Where is the science in EIA? *Search* 20(5): 141–144.

Holling, C.S., editor. 1978. *Adaptive Environmental Assessment and Management*. Toronto: John Wiley & Sons. 377 p.

Lapping, M.B. 1975. Environmental impact assessment methodologies: A critique. *Boston College Environmental Affairs Law Review* 4(1): 123–134.

Leopold, L.B., Clarke, F.E., Hanshaw, B.B., and Balsley, J.R. 1971. *A Procedure for Evaluating Environmental Impact*. Washington, DC: US Department of the Interior. 13 p.

McHarg, I. 1969. *Design With Nature*. Garden City: Natural History Press. 197 p.

Munn, R.E., editor. 1979. *Environmental Impact Assessment: Principles and Procedures*. Second ed. Toronto: John Wiley & Sons. 190 p.

NRC (National Research Council). 1986a. *Ecological Knowledge and Environmental Problem-Solving: Concepts and Case Studies*. Washington, DC: National Academy Press. 388 p.

Rosenberg, D.M., Resh, V.H., Balling, S.S., Barnby, M.A., Collins, J.N., Durbin, D.V., Flynn, T.S., Hart, D.D., Lamberti, G.A., McElravy, E.P., et al. 1981. Recent trends in environmental impact assessment. *Canadian Journal of Fisheries and Aquatic Sciences* 38(5): 591–624.

Shopley, J.B., and Fuggle, R.F. 1984. A comprehensive review of current environmental impact assessment methods and techniques. *Journal of Environmental Management* 18: 25–47.

Sorensen, J.C. 1971. *A framework for the identification and control of resource degradation and conflict in the multiple use of the coastal zone*. Master Thesis. Berkeley: Department of Architecture, University of California. 31 p.

Walters, C.J. 1986. *Adaptive Management of Renewable Resources*. New York: Palgrave Macmillan. 374 p.

Ward, D.V. 1978. *Biological Environmental Impact Studies: Theory and Methods*. New York: Academic Press. 157 p.

Wathern, P. 1984. Methods for assessing indirect impacts. In: *Perspectives on Environmental Impact Assessment*. Clark, B.D., Gilad, A., Bisset, R., et al, editors. Dordecht: Reidel. 213–231 p.

5   Beyond traditional science

5.1 Adaptive management

The concept of adaptive management was originally outlined by Canadian fisheries scientists in the early 1970s (e.g., Walters and Hilborn 1976; Peterman 1977). These scientists observed that complex ecosystems – including the effects of human activities – were shrouded in considerable uncertainty, something that was not being considered in the government regulation of fisheries harvests. These scientists also observed that most fisheries were being modelled and managed based on the assumption that fish stocks – in the absence of human intervention – would remain in a state of static equilibrium. Drawing on the work of ecological theorist Buzz Holling (e.g., Holling 1973), Walters and Hilborn (1976) and Peterman (1977) argued that complex ecosystems are not static, but dynamic, and that they are typically characterized by multiple ecological processes, stability domains, and response thresholds. Overall, these scientists concluded that simulation modelling is a highly useful tool for exploring the future consequences of alternative developments, but that some observation and measurement is ultimately needed to reduce key uncertainties and detect environmental changes. Through the process of adaptive management, such newfound knowledge would then be communicated to managers, regulators, and other stakeholders to inform corrective action (e.g., change harvest rates).

Carl Walters and Ray Hilborn later joined Buzz Holling (1978) and colleagues to outline a framework for adaptive environmental assessment and management. As previously mentioned, the so-called AEAM approach would rely on a series of brief, interdisciplinary workshops and computer simulation modelling exercises to collectively define and evaluate a set of development alternatives based on the predicted responses of environmental variables (i.e., performance indicators). The adaptive management aspect of AEAM would then rely on the monitoring of environmental responses following development to reduce key uncertainties in predictive models and inform subsequent management actions. The long-term management of environmental effects and the ongoing reduction of scientific uncertainty through monitoring are therefore considered to be a single integrated endeavour, requiring close collaboration between researchers and practitioners. While Holling's (1978) book established a strong conceptual basis for AEAM, Walters' (1986) book would later expand on the mathematical basis for adaptive management.

Holling (1978) briefly indicated that "Adaptive management can take a more active form by using the project itself as an experimental probe". Walters and Hilborn (1978) would later elaborate on this statement by defining the terms 'passive' and 'active' adaptive management. According to Walters and Hilborn (1978), passive adaptive management would use "general-processes [sic] studies and previous experience with similar systems to construct the best possible prior model, then manage as if this model were correct while expecting some mistakes that can be used to improve the model as management proceeds". With respect to an active approach, Walters and Hilborn (1978) wrote that active adaptive management treats "all management actions as deliberate experiments that (if properly designed) will have a dual effect of simultaneously producing short-term yields and better information for long-term management." They concluded that passive adaptive management may eventually yield some useful knowledge, but that such an approach wastes time as it does not generate much insight into the behaviour of exploited ecosystems.

In the early 1980s, a major review and evaluation of AEAM (ESSA 1982) found that the approach had been applied in a variety of environmental management contexts since the late 1970s. The report also found, however, that the implementation of AEAM had been limited primarily to the early workshop and modelling stages. In other words, the iterative 'adaptive management' principle of AEAM had yet to be fully realized. As a fundamental barrier to implementation, ESSA (1982) identified the tendency of institutions and even individuals within organizations to resist innovation and change (i.e., adaptation). Perhaps even more fundamentally, ESSA (1982) observed the tendency of individuals and organizations to resist the acknowledgement of uncertainty. The ESSA (1982) report concluded that the implementation of AEAM is as much a political endeavour as a scientific one, requiring the participation of: (i) key individuals within stakeholder organizations to champion and sustain long-term adaptive management programs; (ii) research scientists with relative financial and institutional independence from proponents to focus on reducing uncertainty; and (iii) public stakeholders to ensure transparency in the design, evaluation, and selection of development alternatives.

Despite scientific and political challenges surrounding early implementation efforts, several authors (e.g., Everitt 1983; Jones and Greig 1985; Mulvihill and Keith 1989) continued to praise the AEAM approach during the 1980s, highlighting its applicability to all stages of the general EIA process (e.g., scoping, prediction, monitoring). While Jones and Greig (1985) found that the AEAM approach had yet to be applied throughout any major EIA project to date, they concluded that adaptive management was still highly relevant in addressing many of the key objectives of the EIA process. Later, several authors (e.g., Mulvihill and Keith 1989) would begin to focus more closely on the institutional and organizational arrangements needed to implement adaptive management in the context of EIA.

In his influential book on pursuing sustainable development, Kai Lee (1993) revisited the science of adaptive management, explaining the approach using the metaphor of "a compass: a way to gauge directions when sailing beyond the maps". For Lee (1993), the science of adaptive management would provide a means of reducing critical uncertainties in the pursuit of sustainable development. From a political perspective, Lee (1993) also recognized that conflicts over stakeholder values and perspectives were an increasingly important barrier to achieving mutually agreeable (and sustainable) environmental decisions. Lee (1993) therefore proposed the marriage of adaptive management (Holling 1978; Walters 1986) and the consensus-based politics of principled negotiation. Referring to the politics of negotiation, Lee (1993) wrote: "Democracy, with its contentious stability, is a gyroscope: a way to maintain our bearing through turbulent seas". With respect to both adaptive management and principled negotiation, Lee (1993) concluded that "Compass and gyroscope do not assure safe passage through rough, uncharted waters, but the prudent voyager uses all instruments available, profiting from their individual virtues."

More recently, the literature describes the application of AEAM in a range of environmental management contexts, including several noteworthy successes and failures. Examples include the adaptive management of forests (e.g., Stankey et al. 2003; Bormann et al. 2007), watersheds (e.g., Grayson et al. 1994; Gilmour et al. 1999), rivers (e.g., Meretsky et al. 2000; Susskind et al. 2010), fish (e.g., Hilborn 1992; Smith et al. 1998), wildlife (e.g., Williams et al. 1996; Nichols et al. 2007), wetlands (e.g., Weinstein et al. 1996; Gunderson and Light 2006), and coral reefs (e.g., Hughes et al. 2007; McCook et al. 2010). Whereas some authors (e.g., Walters 1997) have continued to lament the scientific challenges of implementing an adaptive approach to environmental assessment and management, many others (e.g., Lee 1999; Stringer et al. 2006; Allen and Gunderson 2011) have continued to elaborate on the political and institutional challenges of doing so.

While there is ample evidence of AEAM's application within the sphere of renewable resource management, evidence of its implementation in regulatory project-EIA appears to be sparse. Consequently, several authors (e.g., Carpenter 1997; Noble 2000; Thrower 2006; Canter and Atkinson 2010; Benson and Garmestani 2011) have continued to call for its implementation within the context of formal EIA. According to Noble (2000), "Although adaptive management has yet to be fully integrated and tested in the context of EIA, [. . .] principles and applications of adaptive management offer many new possibilities for strengthening the environmental assessment process".

Holling (1978) wrote that "Adaptive management is not really much more than common sense. But common sense is not always in common use". ESSA (1982), however, observed that "There are probably as many definitions of AEAM as there are persons who have been exposed to it". To clarify, ESSA (1982) outlined the three major components of AEAM: (i) concepts of adaptive management, (ii) methods of systems analysis, and (iii) procedures of modelling workshops. To this they added that "Confusion arises when different aspects of these components are emphasized in any one application". In our view, Holling's (1978) framework for AEAM represents a foundational set of principles, protocols, and techniques that *together* provide a strong basis for applying both empirical and deductive biophysical science in EIA.

5.2 Post-normal science

The foundations for what would later become known as post-normal science were laid by philosopher of science Jerry Ravetz in the early 1970s. In his book titled *Scientific Knowledge and Its Social Problems*, Ravetz (1971) argued that modern human societies had thus far benefited tremendously from industrial and technological science, but that such forms of science had also begun to create newer and more challenging problems (e.g., environmental, human health issues). Such problems, he argued, were of a fundamentally different nature than the kinds of problems solved by traditional science. Not only were these new kinds of problems characterized by a high degree of scientific uncertainty, they were imbued with human values and emotions. Ravetz (1971) went on to describe what he saw as the collective search for, and emergence of, a new kind of science that would provide better answers to the seemingly insurmountable social and environmental issues of the day. According to Ravetz (1971), this new, innovative, and much needed mode of inquiry would be called 'critical science'. In explaining the merits of this new and emerging brand of science, Ravetz (1971) wrote that "The work of enquiry is largely futile unless it is followed up by exposure and campaigning; and hence critical science is inevitably and essentially political". Though he observed that the conventional standards of scientific acceptability would be inappropriate for judging the products of this newer kind of science, Ravetz (1971) concluded that "the establishment of criteria for adequacy for solved problems is possible, for the work will frequently be an extension and combination of established fields for new problems, and so critical science can escape the worst perils of immaturity."

In the 1980s, Jerry Ravetz teamed up with philosopher Silvio Funtowicz to elaborate on his new conception of science, this time in terms of 'total environmental assessment'. According to Funtowicz and Ravetz (1985), complex policy-related science could be differentiated into three broad categories – applied science, technical consultancy, and total environmental assessment – based on two dimensions: systems uncertainty and decision stakes. When both dimensions are low, the standard quantitative techniques of applied science would generally suffice. In contrast, when both dimensions are considerable, the quantitative techniques of applied science must be supplemented and interpreted on a qualitative basis by expert judgement. This, according to Funtowicz and Ravetz (1985), was the technical consultancy model of practice. Finally, when decision stakes and systems uncertainty are extremely high, there is a total environmental assessment, "which is permeated by qualitative judgements and value commitments; its result is a contribution to an essentially political debate on larger issues [. . .] The enquiry, even into technical questions, takes the form largely of a dialogue, which may be in an advocacy or even in an adversary mode" (Funtowicz and Ravetz 1985).

Ravetz (1986) later argued that the role of all policy-related science should be to contribute useful knowledge that informs discussions and debates in the political domain. He also argued, however, that what is unknown is often more important than what is known. He proposed that in all kinds of policy-related research, it should also be the role of science to provide participants with 'useful ignorance'. In other words, beyond characterizing what is known, scientific contributions to decision-making should also characterize and communicate what is not known. In cases of total environmental assessment, however, Ravetz (1986) observed that a complete lack of knowledge (i.e., ignorance) precluded the use of traditional statistical tools for uncertainty analysis. Ravetz (1986) further argued that this new kind of research – faced with tremendous scientific and ethical uncertainties – could not realistically be judged against the same criteria as other, more traditional forms of science. He concluded, therefore, that if science and society are to persist in harmony with the biophysical environment, then a new kind of mutual learning would need to take place, one that would require (at minimum) agreed-upon criteria for evaluating the contributions of an emerging science operating in the face of considerable controversy and uncertainty.

Expanding on their framework for understanding and evaluating complex policy-related science, Funtowicz and Ravetz (1990) outlined the so-called NUSAP scheme (Numerical, Unit, Spread, Assessment, Pedigree). Rather than relying on statistical techniques, the NUSAP scheme would use numerical gradations to rank scientific contributions according to both systems uncertainty and information quality. While the NUSAP scheme was designed to accommodate many types of information, Funtowicz and Ravetz (1990) found that "Pedigree [. . .] is the most qualitative and complex of all the categories of the NUSAP notational scheme. Its role is to represent uncertainties that operate at a deeper level than those of the other categories. It conveys an evaluative account of the production process of the quantitative information". The NUSAP scheme – a cornerstone of what would soon become known as post-normal-science – was intended to clearly communicate how different types of uncertainty are related to the quality of scientific information. The end result, it was hoped, would be a shift from the traditional scientific focus on creating 'facts' to an emphasis on exploring the boundary between knowledge and ignorance.

Through a series of papers published in the early 1990s (e.g., Funtowicz and Ravetz 1991, 1992, 1993), the two authors would recast their views on total environmental assessment, eventually settling on the language of 'post-normal science'. Funtowicz and Ravetz (1993) again outlined three kinds of policy-related research based on two dimensions: systems uncertainty and decision stakes. According to Funtowicz and Ravetz (1993), "One way of distinguishing among the different sorts of research is by their goals: applied science is 'mission-oriented'; professional consultancy is 'client-serving'; and post-normal science is 'issue-driven'". Once again, the authors argued that: (i) the uncertainties inherent in post-normal research are not amenable to the statistical techniques of error analysis used in traditional science, and (ii) the quality of contributions made by post-normal science cannot realistically be judged against the same standards as traditional science. According to Funtowicz and Ravetz (1993), post-normal science is "one where facts are uncertain, values in dispute, stakes high, and decisions urgent". To foster quality contributions from this emerging form of science, Funtowicz and Ravetz (1993) argued for the participation of an 'extended peer community', one that would reach beyond the technical and professional specialists involved in traditional peer-review processes. Such widespread participation, they argued, would be needed to build consensus on such unique and complex issues.

Petersen et al. (2011) recently described the experience of implementing a post-normal approach to science in EIA within the Dutch context.

According to Petersen et al. (2011), a scandal in the late 1990s surrounding the management of uncertainty in EIA practice resulted in the Dutch EIA agency adopting guidelines for a post-normal approach to conducting and evaluating scientific contributions. Petersen et al. (2011) explain that the main allegations were that the agency's EIAs "leaned too much towards computer simulation at the expense of measurements" and "suggested too high an accuracy of the environmental figures published". Petersen et al. (2011) also describe two sets of regulatory guidelines developed and issued by the Dutch EIA agency in the early 2000s: one for managing scientific uncertainty and value plurality, and another for stakeholder participation. Petersen et al. (2011) then go on to describe and evaluate the application of the guidelines in several case studies, judging the ability of the guidelines to address three key elements of the post-normal approach: (i) management of uncertainty, (ii) management of multiple perspectives, and (iii) extension of the peer community. Despite evidence of modest progress, the authors conclude that the institutional challenges of implementing practices of post-normal science through a government agency are considerable. In the words of Petersen et al. (2011), "perhaps other institutions are better suited to take the [post-normal science] approach to its extremes".

In our view, the post-normal conception of science represents a strong set of principles for shifting EIA's current focus on creating illusions of 'fact' to openly characterizing, communicating, and reducing key uncertainties. The application of post-normal science in EIA therefore remains entirely consistent and compatible with the application of AEAM.

5.3 Transdisciplinary imagination

In their seminal paper from the early policy science literature, Rittel and Webber (1973) observed the lay public's growing dissatisfaction with traditional academic sciences, particularly when faced with the more intractable social and environmental issues of the day. The authors themselves questioned the ability of traditional scientific disciplines to resolve the unique kinds of problems faced by contemporary public policymakers. Rittel and Webber (1973) observed that unlike the well-bounded or 'tame problems' solved by more traditional and academic forms of science, the more ill-defined or intractable problems faced by planners and policymakers were of a fundamentally different nature. Rittel and Webber (1973) called these problems 'wicked problems'. According to Rittel and Webber (1973), wicked problems can be distinguished from tame problems in that they lack definitive formulations and solutions. In addition to being fundamentally unique, wicked problems are also laden with social values and perspectives, which are often in disagreement with one another. The set of potential solutions to a wicked problem therefore rests largely on one's perception of how that problem is structured.

Rittel and Webber (1973) went on to argue that because resolving a wicked problem is essentially the same process as framing one, policy analysts should not decide on an 'optimal solution' too soon, as is often the case. The underlying paradox here is that the formulation of a wicked problem requires knowledge of potential solutions, which in turn depends on perceptions of problem structure. According to Rittel and Webber (1973), this is the main reason that traditional or academic science – accustomed to neatly defined and agreed-upon problems and solutions – remains incompatible with contemporary wicked problems. The authors also went on to argue that wicked problems, unlike tame problems, are not amenable to scientific experimentation in the classical sense. Because wicked problems are unique, and because each attempt at a solution may cause irreversible social and environmental consequences, such problems cannot be experimented upon without penalty. Though the authors did not provide much guidance for tackling wicked problems, they did hint at a collaborative systems approach to gradually structuring problems and deriving agreeable solutions.

Responding to growing criticism over the role of academic science in modern society, many universities and institutions of higher learning embarked upon programs of educational and organizational reform during the 1970s. Because traditional approaches to knowledge creation, organization, and dissemination were increasingly seen as fragmented and even irrelevant to solving society's most pressing problems, many universities sought to implement arrangements that would foster the conduct of more interdisciplinary research (e.g., Apostel et al. 1972). Still others (e.g., Jantsch 1972; Piaget 1972) advocated a more ambitious transition to what they called 'transdisciplinarity'. According to Jantsch (1972), universities would be increasingly responsible for adopting a new primary purpose: aiding humanity to understand and enhance its collective capacity for self-renewal in the face of rapid social and environmental change. In the words of Jantsch (1972), "The new purpose implies that the university has to become a political institution in the broadest sense, interacting with government (at all jurisdictional levels) and industry in the planning and design of society's systems, and in particular in controlling the outcomes of the introduction of technology into these systems".

Achieving this new purpose, argued Jantsch (1972), would require the design and implementation of new structures for organizing knowledge and conducting research within universities and within larger societies. Jantsch (1972) referred to these new organizational structures – or 'integrated education/innovation systems' – as being transdisciplinary, and outlined a series of steps towards increasing cooperation and coordination between the disciplines. In 'multidisciplinarity', each discipline works within its own traditional boundaries. In 'interdisciplinarity', several disciplines work together in a related manner based on a set of common principles. In 'transdisciplinarity' – the highest level of knowledge integration – research is organized around complex and socially relevant problems that defy traditional disciplinary boundaries. Such transdisciplinary inquiry, argued Jantsch (1972), would be driven by society's most pressing needs, rather than the curiosity of autonomous researchers.

The need to draw upon a variety of disciplines or domains of knowledge in environmental impact studies was originally recognized in the wording of early EIA laws and policies (e.g., Cohen and Warren 1971; White 1972; Cramton and Berg 1973). Despite efforts to bring together multidisciplinary teams of experts and specialists to prepare the first EIAs, early commentaries and reviews (e.g., Carpenter 1976; Schindler 1976; Ward 1978) criticized such attempts for producing little more than fragmented and descriptive reports. To encourage more meaningful communication and collaboration among disciplines, Holling (1978) advocated a series of brief, intense workshops that would require a number of small teams – each comprised of subject matter experts, modellers, and support staff – to rely on the output of one another's modelling efforts as input for their own models. Through a process called 'looking outward', each subsystem modelling group (e.g., hydrology, vegetation, wildlife) would be asked to consider possible links with other subsystems, thereby specifying the information needed from other groups to carry on its own work. Such brief, interdisciplinary workshops, it was hoped, would help to overcome the sprawling and fragmented research that had been the hallmark of EIA practice thus far.

More recently, the notions of wicked problems and transdisciplinarity have seen renewed currency in the context of environmental management and sustainable development (e.g., Jentoft and Chuenpagdee 2009; Hughes et al. 2013; Mauser et al. 2013; Moeliono et al. 2014). In the recent book by Brown et al. (2010) titled *Tackling Wicked Problems through the Transdisciplinary Imagination*, the authors propose the integration and application of both frameworks in pursuing sustainable development. Explaining their choice of the term 'imagination', Brown et al. (2010) write: "imagination is associated with creativity, insight, vision and originality; and it is also related to memory, perception and invention. All of these are necessary in addressing the uncertainty associated with wicked problems in a world of continual change".

Several other authors (e.g., Bond et al. 2010; Greig and Duinker 2011) have recently argued for the application of transdisciplinary working arrangements in the context of formal EIA. According to Bond et al. (2010), current sustainability debates go beyond the questions of multi- or interdisciplinarity posed by earlier attempts at interpreting EIA provisions. Bond et al. (2010) observe that growing demands for public participation and knowledge integration have pushed EIA and related fields of inquiry into the realm of transdisciplinarity. Likewise, the conceptual and organizational framework proposed by Greig and Duinker (2011) for achieving stronger science in EIA is also couched in terms of the transdisciplinary imagination. Here we reiterate the need for transdisciplinary approaches to tackling wicked problems in EIA and observe the complementarity of such an approach with the other frameworks described so far.

5.4 Citizen science

The term 'civic science' has long been used in the literature (e.g., Shen 1975) to mean the exchange of scientific information between researchers and the general public, particularly with respect to information that contributes to better public decision-making. Thus, civic science aims to improve the public's understanding of important scientific knowledge as well as the ability of scientists to communicate that knowledge.

In the 1990s, two natural resource experts – one in Europe (Irwin 1995) and one in North America (Bonney 1996) – began referring to direct public participation in scientific research as 'citizen science'. Each author's use of the term, however, was slightly different. Whereas Bonney (1996) used the term citizen science simply to mean the coordinated collection of environmental data (e.g., bird counts) by members of the lay public (e.g., birdwatchers), Irwin (1995) used it more ambiguously to mean both "science which assists the needs and concerns of citizens" and "a form of science developed and enacted by citizens themselves". In Irwin's (1995) view, a new kind of citizen-driven science would be needed to build confidence in policy-related research and to improve relations between science and society for sustainable development.

In the context of formal EIA, direct public involvement in data collection and interpretation is now commonly referred to as either 'participatory monitoring' or 'community-based monitoring' (e.g., Lawe et al. 2005; Hunsberger et al. 2005; Moyer et al. 2008). Likewise, public involvement in the predictive (i.e., deductive) stage of EIA is now sometimes referred to as 'participatory modelling' (e.g., Videira et al. 2010; Bond et al. 2015). These authors argue that direct involvement of public stakeholders in the scientific stages of EIA ensures greater overall transparency in the process, while also creating more opportunities for stakeholders to influence development decisions and to learn from one another. A number of authors (e.g., Sinclair and Diduck 1995; Sinclair and Diduck 2001; Sinclair et al. 2008) have also argued for the reconception of EIA as a social learning process, one that gives priority to stakeholder education about values, perspectives, environmental impacts, and the EIA process itself. This kind of social learning, it is argued, will inevitably require more informal opportunities for public participation during the design and conduct of EIAs.

Overall, we find that inconsistent use of the terms 'citizen science' and 'civic science' has resulted in confusion surrounding their precise meaning, not to mention how they might be applied in the pursuit of sustainable development (see Clark and Illman 2001). We conclude, however, that the basic notions of science *for* the people (i.e., civic science) and science *by* the people (i.e., citizen science) are both applicable to the implementation of science in EIA and consistent with the other frameworks outlined above.

References

Allen, C.R., and Gunderson, L.H. 2011. Pathology and failure in the design and implementation of adaptive management. *Journal of Environmental Management* 92(5): 1379–1384.

Apostel, L., Berger, G., Briggs, A., and Michaud, G., editors. 1972. *Interdisciplinarity: Problems of Teaching and Research in Universities*. Washington, DC: Organization for Economic Co-operation and Development. 307 p.

Benson, M.H., and Garmestani, A.S. 2011. Embracing panarchy, building resilience and integrating adaptive management through a rebirth of the National Environmental Policy Act. *Journal of Environmental Management* 92(5): 1420–1427.

Bond, A.J., Morrison-Saunders, A., Gunn, J.A.E., Pope, J., and Retief, F. 2015. Managing uncertainty, ambiguity and ignorance in impact assessment by embedding evolutionary resilience, participatory modeling and adaptive management. *Journal of Environmental Management* 151: 97–104.

Bond, A.J., Viegas, C.V., de Souza Reinisch Coelho, C.C., and Selig, P.M. 2010. Informal knowledge processes: The underpinning for sustainability outcomes in EIA? *Journal of Cleaner Production* 18(1): 6–13.

Bonney, R. 1996. Citizen science: A lab tradition. *Living Bird* 15(4): 7.

Bormann, B.T., Haynes, R.W., and Martin, J.R. 2007. Adaptive management of forest ecosystems: Did some rubber hit the road? *Bioscience* 57(2): 186–191.

Brown, V.A., Harris, J.A., Russell, J.Y., editors. 2010. *Tackling Wicked Problems through the Transdisciplinary Imagination*. London: Earthscan.

Canter, L, and Atkinson, S.F. 2010. Adaptive management with integrated decision making: An emerging tool for cumulative effects management. *Impact Assessment and Project Appraisal* 28(4): 287–297.

Carpenter, R.A. 1976. The scientific basis of NEPA – is it adequate? *Environmental Law Report* 6: 50014–50019.

Carpenter, R.A. 1997. The case for continuous monitoring and adaptive management under NEPA. In: *Environmental Policy and NEPA: Past, Present, and Future*. Clark, R., Canter, L., editors. Boca Raton: St. Lucie Press. 163 p.

Clark, F., and Illman, D.L. 2001. Dimensions of civic science: Introductory essay. *Science Communication* 23(1): 5–27.

Cohen, B.S., and Warren, J.M. 1972. Judicial recognition of the substantive requirements of the National Environmental Policy Act of 1969. *Boston College Law Review* 13(4): 685–704.

Cramton, R.C., and Berg, R.K. 1973. On leading a horse to water: NEPA and the federal bureaucracy. *Michigan Law Review* 71(3): 511–536.

ESSA (Environmental and Social Systems Analysts Ltd). 1982. *Review and Evaluation of Adaptive Environmental Assessment and Management.* Ottawa: Environment Canada. 116 p.

Everitt, R.R. 1983. Adaptive environmental assessment and management: Some current applications. In: *Environmental Impact Assessment.* PADC Environmental Impact Assessment and Planning Unit, editor. The Hague: Martinus Nijhoff Publishers. 293–306 p.

Funtowicz, S.O., and Ravetz, J.R. 1985. Three types of risk assessment: A methodological analysis. In: *Environmental Impact Assessment, Technology Assessment, and Risk Analysis: Contributions from the Psychological and Decision Sciences.* Covello, V.T., Mumpower, J.L., Stallen, P.J.M., Uppuluri, V.R.R., editors. Berlin: Springer-Verlag. 831–848 p.

Funtowicz, S.O., and Ravetz, J.R. 1990. *Uncertainty and Quality in Science for Policy.* Dordrecht: Springer Netherlands. 231 p.

Funtowicz, S.O., and Ravetz, J.R. 1991. A new scientific methodology for global environmental issues. In: *Ecological Economics: The Science and Management of Sustainability.* Costanza, R., editor. New York: Columbia University Press. 137–152 p.

Funtowicz, S.O., and Ravetz, J.R. 1992. Three types of risk assessment and the emergence of post-normal science. In: *Social Theories of Risk.* Krimsky, S. and Golding, D., editors. Westport: Greenwood Publishing. 251–273 p.

Funtowicz, S.O., and Ravetz, J.R. 1993. Science for the post-normal age. *Futures* 31(7): 739–755.

Gilmour, A., Walkerden, G., and Scandol, J. 1999. Adaptive management of the water cycle on the urban fringe: Three Australian case studies. *Ecology and Society* 3(1): 11.

Grayson, R.B., Doolan, J.M., and Blake, T. 1994. Application of AEAM (adaptive environmental assessment and management) to water quality in the Latrobe River catchment. *Journal of Environmental Management* 41(3): 245–258.

Greig, L.A., and Duinker, P.N. 2011. A proposal for further strengthening science in environmental impact assessment in Canada. *Impact Assessment Project Appraisal* 29(2): 159–165.

Gunderson, L., and Light, S.S. 2006. Adaptive management and adaptive governance in the Everglades ecosystem. *Policy Sciences* 39(4): 323–334.

Hilborn, R. 1992. Institutional learning and spawning channels for sockeye salmon (*Oncorhynchus nerka*). *Canadian Journal of Fisheries and Aquatic Sciences* 49(6): 1126–1136.

Holling, C.S. 1973. Resilience and stability of ecological systems. *Annual Review of Ecology Systematics* 4: 1–23.

Holling, C.S., editor. 1978. *Adaptive Environmental Assessment and Management*. Toronto: John Wiley & Sons. 377 p.

Hughes, T.P., Gunderson, L.H., Folke, C., Baird, A.H., Bellwood, D., Berkes, F., Crona, B., Helfgott, A., Leslie, H., Norberg, J., et al. 2007. Adaptive management of the Great Barrier Reef and the Grand Canyon World Heritage Areas. *Ambio* 36(7): 586–592.

Hughes, T.P., Huang, H., and Young, M.A.L. 2013. The wicked problem of China's disappearing coral reefs. *Conservation Biology* 27(2): 261–269.

Hunsberger, C.A., Gibson, R.B., and Wismer, S.K. 2005. Citizen involvement in sustainability-centred environmental assessment follow-up. *Environmental Impact Assessment Review* 25(6): 609–627.

Irwin, A. 1995. *Citizen Science: A Study of People, Expertise and Sustainable Development*. Abingdon: Routledge. 216 p.

Jantsch, E. 1972. Towards interdisciplinarity and transdisciplinarity in education and innovation. In: *Interdisciplinarity: Problems of Teaching and Research at Universities*. Apostel, L., Berger, G., Briggs, A., et al, editors. Washington, DC: Organization for Economic Co-operation and Development. 97–121 p.

Jentoft, S., and Chuenpagdee, R. 2009. Fisheries and coastal governance as a wicked problem. *Mar Policy* 33(4): 553–560.

Jones, M.L., and Greig, L.A. 1985. Adaptive environmental assessment and management: A new approach to environmental impact assessment. In: *New Directions in Environmental Impact Assessment in Canada*. Maclaren, V.W. and Whitney, J.B., editors. Toronto: Methuen Publications. 21–42 p.

Lawe, L.B., Wells, J., and Cree, M. 2005. Cumulative effects assessment and EIA follow-up: A proposed community-based monitoring program in the Oil Sands Region, Northeastern Alberta. *Impact Assessment Project Appraisal* 23(3): 205–209.

Lee, K.N. 1993. *Compass and Gyroscope: Integrating Science and Politics for the Environment*. Washington, DC: Island Press. 243 p.

Lee, K.N. 1999. Appraising adaptive management. *Ecology and Society* 3(2): 3.

Mauser, W., Klepper, G., Rice, M., Schmalzbauer, B.S., Hackmann, H., Leemans, R., and Moore, H. 2013. Transdisciplinary global change research: The co-creation of knowledge for sustainability. *Current Opinion in Environmental Sustainability* 5(3–4): 420–431.

McCook, L.J., Ayling, T., Cappo, M., Choat, J.H., Evans, R.D., Freitas, D.M.D., Heupel, M., Hughes, T.P., Jones, G.P., Mapstone, B., et al. 2010. Adaptive management of the great barrier reef: A globally significant demonstration of the benefits of networks of marine reserves. *Proceedings of the National Academy of Sciences of USA* 107(43): 18278–18285.

Meretsky, V.J., Wegner, D.L., and Stevens, L.E. 2000. Balancing endangered species and ecosystems: A case study of adaptive management in Grand Canyon. *Environmental Management* 25(6): 579–586.

38   *Beyond traditional science*

Moeliono, M., Gallemore, C., Santoso, L., Brockhaus, M., and Di Gregorio, M. 2014. Information networks and power: Confronting the "wicked problem" of REDD+ in Indonesia. *Ecology and Society* 19(2): 9.

Moyer, J., Fitzpatrick, P., Diduck, A., and Froese, B. 2008. Towards community-based monitoring in Manitoba's hog industry. *Canadian Public Administration* 51(4): 637–658.

Mulvihill, P.R., and Keith, R.F. 1989. Institutional requirements for adaptive EIA: The Kativik Environmental Quality Commission. *Environmental Impact Assessment Review* 9(4): 399–412.

Nichols, J.D., Runge, M.C., Johnson, F.A., and Williams, B.K. 2007. Adaptive harvest management of North American waterfowl populations: A brief history and future prospects. *Journal of Ornithology* 148(2): 343–349.

Noble, B.F. 2000. Strengthening EIA through adaptive management: A systems perspective. *Environmental Impact Assessment Review* 20(1): 97–111.

Peterman, R.M. 1977. A simple mechanism that causes collapsing stability regions in exploited salmonid populations. *Journal of Fisheries Research Board of Canada* 34(8): 1130–1142.

Petersen, A.C., Cath, A., Hage, M., Kunseler, E., and van der Sluijs, J.P. 2011. Postnormal science in practice at the Netherlands environmental assessment agency. *Science Technology & Human Values* 36(3): 362–388.

Piaget, J. 1972. The epistemology of interdisciplinary relationships. In: *Interdisciplinarity: Problems of Teaching and Research in Universities*. Apostel, L., Berger, G., Briggs, A., et al, editors. Washington, DC: Organization for Economic Co-operation and Development. 127–139 p.

Ravetz, J. 1971. *Scientific Knowledge and its Social Problems*. Oxford: Oxford University Press. 449 p.

Ravetz, J.R. 1986. Usable knowledge, usable ignorance: Incomplete science with policy implications. In: *Sustainable Development of the Biosphere*. Clark, W.C. and Munn, R.E., editors. Cambridge: Cambridge University Press. 415–432 p.

Rittel, H.W.J., and Webber, M.M. 1973. Dilemmas in a general theory of planning. *Policy Sciences* 4(2): 155–169.

Schindler, D.W. 1976. The impact statement boondoggle. *Science* 192(4239): 509.

Shen, B.S. 1975. Science literacy and the public understanding of science. In: *Communication of Scientific Information*. Day, S.B., editor. Basel: Karger Publishers. 44–52 p.

Sinclair, A.J., and Diduck, A. 1995. Public education: An undervalued component of the environmental assessment public involvement process. *Environmental Impact Assessment Review* 15(3): 219–240.

Sinclair, A.J., Diduck, A., and Fitzpatrick, P. 2008. Conceptualizing learning for sustainability through environmental assessment: Critical reflections on 15 years of research. *Environmental Impact Assessment Review* 28(7): 415–428.

Sinclair, A.J., and Diduck, A.P. 2001. Public involvement in EA in Canada: A transformative learning perspective. *Environmental Impact Assessment Review* 21(2): 113–136.

Smith, C.L., Gilden, J., Steel, B.S., and Mrakovcich, K. 1998. Sailing the shoals of adaptive management: The case of salmon in the Pacific Northwest. *Environmental Management* 22(5): 671–681.

Stankey, G.H., Bormann, B.T., Ryan, C., Shindler, B., Sturtevant, V., Clark, R.N., and Philpot, C. 2003. Adaptive management and the Northwest Forest Plan: rhetoric and reality. *Journal of Forestry* 101(1): 40–46.

Stringer, L.C., Dougill, A.J., Fraser, E., Hubacek, K., Prell, C., and Reed, M.S. 2006. Unpacking "participation" in the adaptive management of social – ecological systems: A critical review. *Ecology and Society* 11(2): 39.

Susskind, L., Camacho, A.E., and Schenk, T. 2010. Collaborative planning and adaptive management in Glen Canyon: A cautionary tale. *Columbia Journal of Environmental Law* 35: 1–56.

Thrower, J. 2006. Adaptive management and NEPA: How a non-equilibrium view of ecosystem mandates flexible regulation. *Ecology Law Quarterly* 33: 871–896.

Videira, N., Antunes, P., Santos, R., and Lopes, R. 2010. A participatory modeling approach to support integrated sustainability assessment processes. *Systems Research and Behavioral Science* 27(4): 446–460.

Walters, C. 1997. Challenges in adaptive management of riparian and coastal ecosystems. *Ecology and Society* 1(2): 1.

Walters, C.J. 1986. *Adaptive Management of Renewable Resources.* New York: Palgrave Macmillan. 374 p.

Walters, C.J., and Hilborn, R. 1976. Adaptive control of fishing systems. *Journal of the Fisheries Research Board of Canada* 33(1): 145–159.

Walters, C.J., and Hilborn, R. 1978. Ecological optimization and adaptive management. *Annual Review Ecology and Systematics* 9: 157–188.

Ward, D.V. 1978. *Biological Environmental Impact Studies: Theory and Methods.* New York: Academic Press. 157 p.

Weinstein, M.P., Balletto, J.H., Teal, J.M., and Ludwig, D.F. 1996. Success criteria and adaptive management for a large-scale wetland restoration project. *Wetlands Ecology and Management* 4(2): 111–127.

White, G.F. 1972. Environmental impact statements. *The Professional Geographer* 24(4): 302–309.

Williams, B.K., Johnson, F.A., and Wilkins, K. 1996. Uncertainty and the adaptive management of waterfowl harvests. *Journal of Wildlife Management* 60(2): 223–232.

6 Emerging concepts for science in EIA

6.1 Resilience

The concept of resilience in ecological systems was first articulated by Canadian ecologist Buzz Holling. According to Holling (1973), the traditional equilibrium-centred view of ecosystems – inherited from the physical sciences – could not adequately explain how ecosystems actually responded to natural and anthropogenic disturbances. Rather than simply returning to a single stable equilibrium, many ecosystems seemed to remain in an altered or degraded state, despite removal or cessation of the initial disturbance. Using two- and three-dimensional phase plane diagrams, Holling (1973) showed how complex ecosystems may in fact be characterized by multiple domains of stability, the shape of each being determined by a particular set of parameter values. Whereas the equilibrium-centred view of ecosystems focused on the degree of constancy within a narrow band of stable (i.e., predictable) conditions, the resilience view of ecosystems would focus on the boundaries (i.e., thresholds) separating multiple domains of attraction, each with its own unique stability characteristics.

Holling (1973) wrote that: "Resilience determines the persistence of relationships within a system and is a measure of the ability of these systems to absorb changes of state variables, driving variables, and parameters [. . .] Stability, on the other hand, is the ability of a system to return to an equilibrium state after a temporary disturbance". With respect to diversity, stability, and resilience, Holling (1973) concluded that – contrary to traditional thinking – ecosystems with more diversity (i.e., relationships) were in fact *less* stable than ecosystems with fewer relationships. He argued that it was this dynamic variability – the product of a long and stressful evolutionary history – that ultimately made such ecosystems so resilient.

With respect to resource management, Holling (1973) argued that equilibrium-centred approaches, which aim to reduce the natural variability of ecosystems (e.g., flood, fire control), actually serve to erode their resilience to change, slowly rendering them fragile and vulnerable to sudden collapse. Under such restrictive management regimes, a slow quantitative change in parameter values, with no accompanying qualitative response, might suddenly give way to a dramatic flip into another stable ecosystem configuration from which there is little hope of return. A resilience-based approach to ecosystem management, as proposed by Holling (1973), would instead focus on the need to maintain the spatial heterogeneity and temporal variability inherent in natural ecosystems, thereby preserving their ability to absorb future events and persist in the face of unexpected change.

As part of the emerging AEAM framework, Holling (1978) expanded on the notion of resilience to include the sociopolitical dimension. Holling (1978) proposed that human individuals, societies, and institutions – the counterparts of ecological systems and components – also exhibited properties of stability and resilience. According to Holling (1978), many large institutions, such as government agencies, businesses, and universities, were also being managed with the aim of reducing variability and controlling exposure to disturbance. Holling (1978) argued that all of this was done at the cost of maintaining responsiveness and adaptability in the face of unexpected change.

With growing concerns over the consequences of global climate change, Holling (1986) expanded on the notion of ecological resilience, using it to explain the relationship between local ecosystem complexity, global biogeochemical cycles, and sustainable development. First, he began by observing that vegetation succession, an important concept in understanding the dynamics of terrestrial ecosystems, had thus far only provided a partial picture of ecosystem dynamics. The prevailing theory of 'old field succession' had been conceived of simply as: (i) the rapid exploitation of abiotic resources by fast-growing pioneer species following a disturbance, and (ii) the gradual establishment and maturation of more competitive but slower growing climax species. To better understand the basic dynamics of ecosystems, Holling (1986) proposed a figure-eight–shaped cycle representing the sequential interaction of four aggregate ecosystem processes: (i) exploitation, (ii) conservation, (iii) creative destruction, and (iv) renewal. During the exploitation and conservation stages (the focus of traditional successional theory), ecosystems rapidly consolidate and organize matter and energy as the degree of connectedness between species and components increases. Eventually, things become too connected, homogeneous, and self-similar, and some kind of random disturbance event triggers the onset of creative destruction. Here, stored capital is suddenly released, only to be exploited and reorganized all over again. Though Holling (1986) initially called this the ecosystem cycle, it would later become known as the adaptive cycle.

According to Holling (1986), the dynamics of creative destruction and renewal play an important role in shaping the stability and resilience properties of ecosystems, for it is here that materials and energy are made available for the creative forces of renewal. During these brief windows of opportunity, ecosystems are able to break old patterns and structures, reorganize, and adapt to evolving internal and external forces of change. It is also during these sudden releases, however, that ecosystems risk the irretrievable loss of material, forever altering their stability and resilience properties.

More recently, the notion of ecological resilience has been elaborated upon in the context of other emerging concepts such as complexity (e.g., Gunderson and Holling 2001; Holling 2001; Puettmann et al. 2009), biodiversity (e.g., Peterson et al. 1998; Folke et al. 2004), and sustainability (e.g., Arrow et al. 1995; Folke et al. 2002). Making the connection between biodiversity and ecosystem resilience, Peterson et al. (1998) concluded that "ecological resilience is generated by diverse, but overlapping, function within a scale and by apparently redundant species that operate at different scales. The distribution of functional diversity within and across scales allows regeneration and renewal to occur following ecological disruption". Likewise, relating the notion of ecosystem resilience to that of sustainable development, Arrow et al. (1995) concluded that "If human activities are to be sustainable, we need to ensure that the ecological systems on which our economies depend are resilient". Simply put, sustainable development relies on the maintenance of ecological resilience, which in turn relies on the preservation of biodiversity.

At the same time, other authors (e.g., Gunderson 1999) have continued to elaborate on the relationship between resilience and adaptive management. According to Gunderson (1999), many of the observed failures of the AEAM approach can be attributed to a fundamental lack of resilience and flexibility in natural ecosystems, as well as in human societies. In the words of Gunderson (1999), "the successes and failures of AEAM are intertwined with system properties of flexibility and resilience. In a nutshell, if there is no resilience in the ecological system, nor flexibility among stakeholders in the coupled social system, then one simply cannot manage adaptively". In such cases, Gunderson (1999) concluded, efforts must be made to restore social and ecological resilience before adaptive management may successfully proceed.

The importance of adopting a resilience perspective in the context of formal EIA has also been highlighted recently in the literature, both in terms of project-level (e.g., Benson and Garmenstani 2011; Bond et al. 2015) and strategic-level EIAs (e.g., Slootweg and Jones 2011). From an administrative perspective, Benson and Garmestani (2011) argue that current EIA laws and policies, if reconfigured, could provide an appropriate regulatory home for adaptive management. The authors also observe, however, that existing environmental laws and regulations are typically rigid, having been established during a time when the equilibrium-centered view of ecosystems prevailed. To successfully build resilience into ecological systems (and human institutions), Benson and Garmenstani (2011) argue for a reconfiguration and rewording of EIA laws and policies to reflect the ideals of an iterative and adaptive approach to environmental assessment and management. From both scientific and political perspectives, Bond et al. (2015) highlight the importance of resilience thinking in EIA, both in terms of ecological resilience as well as embedding evolutionary resilience within the EIA process itself. In the words of Bond et al. (2015), "uncertainty, ambiguity and ignorance [. . .] can be better managed through embedding an evolutionary resilience approach, supported through participatory deliberation and adaptive management". Here we echo Bond et al.'s (2015) proposal for "embedding evolutionary resilience, participatory modelling and adaptive management" into formal EIA practice. We also reiterate the urgent need to translate, test, and refine existing theory on the subject.

6.2 Thresholds

The concept of response thresholds has had a long history in ecology, particularly with respect to the study of predator-prey interactions (e.g., Gause et al. 1936; Holling 1966; Larkin et al. 1964; Ware 1972). The term was originally used by population/community ecologists to mean an abrupt, non-linear, and potentially irreversible change in the abundance or behaviour of animal populations within a particular predator-prey system. Similar concepts like assimilative capacity (e.g., Cairns 1977) and critical loads (e.g., Wong and Clark 1976) were used to conceptualize the ability of aquatic ecosystems to absorb or tolerate releases of anthropogenic pollutants. In theory, significant or irreversible environmental effects would be likely to occur if such limitations were exceeded.

Holling (1973) famously applied the notion of ecological thresholds to explain the general stability properties of complex ecosystems. According to Holling (1973), such complex ecosystems are typically characterized by multiple equilibria, each situated within its own unique stability domain. Holling (1973) conceived of such stable ecosystem states as being separated by theoretical boundaries called escape thresholds. Should changes in ecosystem state variables or parameters exceed such thresholds, either due to planned human intervention or stochastic disturbance events, an ecosystem could potentially flip into an alternate state of being from which it may be difficult or even impossible to return.

Holling (1986) later argued that the existence of ecological stability domains, as well as the boundaries (i.e., thresholds) separating them, could be attributed to a small number of ecological processes operating at different spatiotemporal scales, and hence at different speeds. Such ecological stability domains (and the thresholds separating them) could be defined by the interaction of fast, intermediate, and slow variables nested within a spatial hierarchy. According to Holling (1986), it was the interaction of these three distinct speeds of variables that produced the cyclic variations typically observed in most ecosystems.

With respect to formal EIA, the notion of ecological thresholds has been an important concept in determining the significance of environmental impacts since the 1980s. Conover et al. (1985) outlined a well-known approach to determining the significance of environmental impacts based on the application of ecologically based thresholds. Such quantitative limits were intended to determine the assignment of qualitative significance rankings (e.g., major, moderate, minor, negligible) to predicted environmental impacts. Still other authors (e.g., Sassaman 1981; Haug et al. 1984) outlined somewhat more political approaches to impact significance determination based on the application of socially derived 'thresholds of concern'.

The notion of applying ecological thresholds to protect environmental values from the unwanted effects of human development has recently received increasing attention in the literature. While some authors (e.g., Brenden et al. 2008; Sorensen et al. 2008; Andersen et al. 2009; Ficetola and Denoël 2009) have outlined quantitative techniques for identifying and delineating ecological thresholds, others (e.g., Scheffer and Carpenter 2003; Bestelmeyer 2006; Groffman et al. 2006) have lamented the practical difficulties of doing so. Still others (e.g., Johnson 2013) have explored the fundamental differences and similarities between ecological, social, and regulatory thresholds. With respect to ecological thresholds and environmental decision-making, Johnson (2013) concluded that "Scientists [. . .] will play the lead role in providing the technical information that is the ecological threshold, including a full accounting of uncertainty in such information and some perspective on the implications of conservation decisions that exceed those thresholds".

The use of thresholds for determining environmental impact significance has continued to receive considerable attention in peer-reviewed EIA literature. While some authors (e.g., Duinker and Greig 2006; Duinker et al. 2013) have advocated the use of ecologically based thresholds for determining the significance of environmental impacts, others (e.g., Canter and Canty 1993; Ehrlich and Ross 2015) have advocated the use of socially derived thresholds of acceptability. Still others (e.g., Kjellerup 1999; Schmidt et al. 2008) have advocated the use of regulatory standards and thresholds for determining the significance of environmental impacts. Without a doubt, ecological response thresholds – if they can be identified – represent the maximum degree to which ecosystem components or processes can be altered and still deemed sustainable. We argue, therefore, that such inherent biophysical limitations must provide the basis for all political and regulatory discussions surrounding environmental impact significance.

6.3 Complexity

The notion of complexity has had a long history in ecology. Here we cite early debates surrounding the so-called complexity-stability hypothesis (e.g., MacArthur 1955; Elton 1958). Such arguments postulated that increases in trophic web complexity would lead to increases in community stability. This, it was argued, was due to greater redundancy of ecosystem functions. In the event that one trophic linkage was lost, many others would continue to provide that particular function, effectively buffering change and ensuring ecosystem stability. Other authors (e.g., May 1973; Holling 1973), however, challenged this traditional view of ecosystems, arguing that increases in ecosystem complexity would actually lead to greater ecosystem *in*stability.

In Holling's (1973) view, complex ecosystems could be mathematically reduced to a set of key state variables governed by changes in a related set of driving variables. The form and function of the relationship connecting such variables could then be determined by estimating a set of parameters. According to Holling (1973), it is the number, size, and shape of these functional stability domains that determine the overall robustness (i.e., resilience) of ecosystems. In explaining the general characteristics of complex ecosystems, Holling (1973) pointed to multiple stable states, feedback loops, nonlinear thresholds, response lags, and density-dependent relationships. Holling (1973) concluded that it is precisely the unstable nature of complex ecosystems that allows them to persist and adapt in a changing world.

While Holling (1973) saw complex ecosystems as being fundamentally unstable and unpredictable, he also viewed their dynamic behaviour as being organized around stable domains of attraction. May (1974), on the other hand, argued that the behaviour of complex ecosystems is often highly disorganized or even chaotic. In cases of chaos, May (1974) argued, ecosystems may be extremely sensitive to starting conditions, meaning that slight differences in input variables can result in vastly divergent and fundamentally unpredictable outcomes. This perspective on the behaviour of complex systems has since been dubbed 'chaos theory'.

Holling (1986) later expanded on his theory of complex ecosystems, this time emphasizing the hierarchical or cross-scale connections linking ecosystems operating at vastly different spatial and temporal scales. According to Holling (1986), local ecosystems could be best understood as being embedded within earth's global biogeochemical cycles. While the complexity of local ecosystems could still be captured mathematically in terms of state variables, driving variables, and parameters, driving variables in particular would have to be differentiated in terms of their internal or external orientation to the ecosystem in question. In this way, broad-scale phenomena such as global climate change could be understood as the emergent effect of many small changes occurring at the local level. Conversely, the effect of global climate change on local ecosystems could also be seen as a kind of cross-scale feedback loop.

In addition to addressing issues of scale, hierarchy, emergence, and feedback in complex ecosystems, Holling (1986) sought to provide a synthesis of organized and disorganized notions of complexity. To this end, Holling (1986) introduced what he dubbed the 'ecosystem cycle' (later referred to as the adaptive cycle) to explain how complex ecosystems tend to alternate between periods of gradually increasing organization (i.e., renewal, exploitation, conservation) and periods of sudden disorganization (i.e., creative destruction).

More recently, several authors (e.g., Gunderson and Holling 2001; Holling 2001) have elaborated on the relationships linking diversity, stability, productivity, and resilience in complex ecosystems. These authors have also extended their perspectives on resilience and complexity to include social and economic systems as well as ecological ones. Arguing that the term 'hierarchy' has become "so burdened by the rigid, top-down nature of its common meaning", Holling (2001) "decided to look for another term that would capture the adaptive and evolutionary nature of adaptive cycles that are nested one within each other across space and time scales". Ultimately, he would call such complex, adaptive, multi-scale systems 'panarchies'. In the words of Holling (2001), "A panarchy is a representation of a hierarchy as a nested set of adaptive cycles. The functioning of those cycles and the communication between them determines the sustainability of a system". Whereas the term 'hierarchy' implies top-down control of lower levels by higher levels, the term 'panarchy' implies that all levels or subsystems can influence one another.

From a somewhat more practical perspective, several authors (e.g., Kay et al. 1999; Pahl-Wostl 2007; Puettmann et al. 2009) have outlined frameworks for explicitly managing forests and other renewable resources as complex adaptive systems. According to Puettmann et al. (2009), the overall aim of such an approach is to preserve ecosystem resilience and adaptive capacity, for it is the complexity and diversity of ecosystems that allows them to reorganize, innovate, and adapt in the face of sudden change. The maintenance of ecosystem complexity is aimed at preserving the ability of ecosystems to evolve and persist into the distant future, a prerequisite for achieving sustainable development. For general understanding, Puettmann et al. (2009) outline six essential attributes of complex ecosystems: (i) non-linear relationships, (ii) ill-defined boundaries, (iii) disequilibrium, (iv) self-organizing feedback loops, (v) emergent behaviours, and (vi) memory of past states. Puettmann et al. (2009) conclude with four key principles for managing forests and other renewable resources as complex adaptive systems: (i) consider a variety of ecosystem components; (ii) accept natural variability in space and time; (iii) actively maintain and develop heterogeneity in ecosystem structure, composition, and function; and (iv) predict and measure success at multiple scales.

With respect to the general applicability of panarchy theory to pursuing sustainable development, Holling (2001) concluded that "The theory is sufficiently new that its practical application to regional questions or the analysis of specific problems has just begun". By offering a means of simplifying complexity while simultaneously capturing its essential elements, we feel that complexity theory provides a powerful framework for understanding the dynamic behaviour of ecosystems in the context of EIA.

6.4 Landscape ecology

The field of landscape ecology as we know it today emerged during the 1980s, a time when ecological theorists were grappling with the confounding spatial dimension of complex ecosystems. The first landscape ecologists (e.g., Forman and Godron 1981; Forman and Godron 1986) sought to describe the different kinds of spatial structures and patterns commonly found in natural landscapes (e.g., patch, corridor, matrix). Taking things a step further, other authors (e.g., Turner 1989) sought to understand the relationships linking such spatial patterns and structures to ecological processes operating at somewhat finer scales (e.g., wildlife population dynamics). Similarly, other authors (e.g., Shugart and West 1981; Urban et al. 1987) sought to understand the relationships linking scale, natural disturbance regimes, and spatial vegetation patterns within dynamic landscapes. Still other authors (e.g., Burgess and Sharpe 1981; Naveh 1982) sought to understand the effects of human development actions on particular landscape patterns and structures.

At the time of EIA's genesis in the 1970s, landscape ecology was in its infancy (e.g., Troll 1971; Levin and Paine 1974; Wiens 1976; Hansson 1977). By the time of the Beanlands and Duinker (1983) study, the field was just beginning to flourish (e.g., Forman and Godron 1981; Burgess and Sharpe 1981). Although the establishment of formal EIA processes and the development of landscape ecology occurred virtually concurrently, the emerging concepts and techniques of landscape ecology had not yet found major application in mainstream EIA practice. At this time, other domains of ecology were still prominent in EIA, such as the analysis of food webs (e.g., Karr 1981; Hecky et al. 1984), nutrient cycles (e.g., Hecky et al. 1984), and wildlife habitat (e.g., Severinghaus 1981; Landres 1983; Hecky et al. 1984).

Later in the 1980s, a number of scientific guidance materials (e.g., Bedford and Preston 1988; Lee and Gosselink 1988; Weller 1988; Winter 1988) established the applicability of landscape ecology to assessing the cumulative effects of human development on the spatial structure, function, and diversity of terrestrial ecosystems, particularly forests and wetlands. These authors argued that predicting and measuring the environmental impacts of human development at the local or site scale was simply not enough to infer the potential for cumulative effects of multiple developments occurring over much broader scales.

In a related set of developments, the first spatially explicit landscape simulation models were created to predict the effects of human development on wetland and forest ecosystems (e.g., Kessell et al. 1984; Pearlstine et al. 1985; Costanza et al. 1988; Sklar et al. 1985). Since then, more sophisticated simulation models have been developed to predict the effects of timber harvesting, natural disturbance, and climate change on the structure, composition, diversity, and productivity of managed forest landscapes (e.g., Mladenoff and He 1999; Scheller and Mladenoff 2004; Scheller et al. 2007). Such models have found widespread application in the management of forests (e.g., Gustafson et al. 2000) and the conservation of threatened wildlife species (e.g., Akçakaya 2001; Larson et al. 2004).

Gontier et al. (2006) explored the possibility of using landscape simulation models in the context of formal EIA. From a biodiversity perspective, the authors proposed the use of such models during the predictive and evaluative stages of EIA processes. With respect to empirical research, several recent studies have used biotelemetry techniques to quantify the effects of linear development structures (e.g., roads, pipelines, seismic lines) on habitat use by terrestrial mammals, particularly wolves and caribou (e.g., Houle et al. 2010; Sorensen et al. 2008).

In our view, these are welcome developments, as they demonstrate how landscape ecology can be used as a powerful set of lenses for understanding ecosystemic responses to development. We therefore expect to see the tools and methods of landscape ecology used prominently in ecological impact assessments in the near future. One potential reason for an imperfect fit, however, is the notion that EIA is driven by stakeholder-relevant VECs that are normally identified on the basis of traditional elements of ecosystems (e.g., particular species, specific ecosystems, air and water quality). Indeed, it would seem that the concepts of landscape ecology (e.g., connectivity, patch, matrix) are not yet in the normal lexicon of most EIA participants. We therefore hypothesize that landscape ecology may well be serving EIA studies in the background but may not be found prominently in the documents directly associated with EIAs.

6.5 Biodiversity

Modern notions of biodiversity can be traced back to earlier concepts like species diversity and species richness (e.g., Hurlbert 1971; Whittaker 1972; Peet 1974; Huston 1979). In the 1980s, many ecologists (e.g., Wilson and Peter 1988) began using the terms 'biodiversity' and 'biological diversity' to mean the variety of life at all levels of biological organization. While many authors (e.g., Myers 1988; Lugo 1988) continued to use the term biodiversity to mean simple species diversity, others (e.g., Franklin 1988; Ray 1988) began using it more broadly to mean diversity of genetic material, species, functional groups, ecosystems, habitats, and landscape structures.

To quantify and protect biodiversity at multiple levels of organization, some authors (e.g., Noss 1987; Hunter et al. 1988) advocated a so-called coarse-filter/fine-filter approach. Using this approach, a coarse-filter – representing a variety of important landscape structures, ecosystem types, community assemblages, and successional stages – is first applied to capture biodiversity in the broadest sense. Next, a fine-filter – comprised of habitat requirements for particular species – is applied to capture any rare or unevenly distributed species that may have slipped through the coarse-filter. In the words of Hunter et al. (1988): "The conservation of biological diversity is too complex for monolithic approaches; and preservation of populations, species, communities, physical environments, ecosystems, and landscapes must all be considered, when deemed appropriate".

In the early 1990s, the United Nations Earth Summit in Rio named the 'conservation of biological diversity' a major precondition for achieving long-term ecological resilience and sustainable development. According to Agenda 21 (UNCED 1992): "Urgent and decisive action is needed to conserve and maintain genes, species and ecosystems, with a view to the sustainable management and use of biological resources. Capacities for the assessment, study and systematic observation and evaluation of biodiversity need to be reinforced". The report went on explicitly to identify formal project EIA as a means of protecting Earth's biodiversity from any unwanted effects of human development.

Since then, an abundance of scholarly articles and guidance materials have outlined frameworks for integrating biodiversity considerations into formal EIA practice (e.g., Nelson and Serafin 1991; Nelson and Serafin 1992; CEQ 1993; CEAA 1996; Diaz et al. 2001; Slootweg and Kolhoff 2003; Geneletti 2003; Mandelik et al. 2005; Gontier et al. 2006; Wale and Yalew 2010; Slootweg et al. 2010). Recognizing a plurality of perspectives on the subject, Nelson and Serafin (1991) outlined a participatory approach to biodiversity assessment in which participants are invited to identify and discuss important components of biological diversity within the local or regional context. Such an approach to assessment, they argue, helps to ensure the production of scientific information that is useful and relevant to all EIA participants. Still other authors (e.g., Gontier et al. 2006; Mörtberg et al. 2007) have explored the possibility of applying the tools and techniques of landscape ecology to predict the effects of human development on components of terrestrial biodiversity.

Despite calls for the application of biodiversity concepts in formal EIA, reviews of practice (e.g., Mandelik et al. 2005; Wegner et al. 2005; Gontier et al. 2006; Söderman 2006; Khera and Kumar 2010) have generally found the inclusion of such elements to be weak. As with the principles of landscape ecology, we suspect this may reflect an imperfect fit between assessments based on readily identifiable elements of ecosystems (e.g., species, air and water quality) and assessments based on abstract components of biodiversity (e.g., ecosystem types, habitat types, patch connectivity). Despite such differences, we expect to see the various components of biodiversity increasingly considered alongside more traditional ecosystem components in mainstream EIA practice. In our view, the inclusion of biodiversity considerations in EIA will help to protect the overall resilience of ecosystems and, ultimately, the sustainability of VECs.

6.6 Sustainability

The notion of sustainable development was popularized in the late 1980s by the United Nations World Commission on Environment and Development (WCED 1987), which defined it as "development that meets the needs of the present without compromising the ability of further generations to meet their own needs". The WCED (1987) report went on to describe sustainable development as being: (i) contained within ecological limits and (ii) socially agreeable and equitable. Simply put, sustainable development is that which conserves valuable environmental resources for continued use by present and future generations. The WCED (1987) also went on to explicitly identify formal EIA as a means of pursuing sustainable development, arguing for the expansion of EIA processes to consider the environmental effects of policies and programs as well as individual projects. According to the WCED (1987) report, "This broader environmental assessment should be applied not only to products and projects, but also to policies and programmes, especially major macroeconomic, finance, and sectoral policies that induce significant impacts on the environment".

Shortly after the WCED's (1987) declaration on sustainability, an abundance of papers was published on the role of formal EIA in achieving sustainable development (e.g., Rees 1988; Gardner 1989; Jacobs and Sadler 1990).

Drawing on the Beanlands and Duinker (1983) report, Gardner (1990) wrote: "Even traditional EIA, especially when strengthened and focused by the 'ecological framework', plays an essential role in fostering awareness of the environmental consequences of development activities and of tradeoffs involved".

Since then, the notion of EIA as a tool for pursuing sustainable development has been widely endorsed by scholars and practitioners alike (e.g., Sadler 1996; Lawrence 1997; Sinclair et al. 2008). At the same time, other authors (e.g., Pope et al. 2004; Gibson et al. 2005; Gibson 2006; Weaver and Rotmans 2006) have argued for a more 'integrated' approach to evaluating the social and environmental effects of proposed developments. These authors collectively refer to this approach as 'sustainability assessment'. Gibson (2006), for example, has proposed a set of 'core generic criteria' and 'trade-off rules' – encompassing multiple dimensions of sustainability (e.g., social, environmental, economic) – to serve as a framework for assessment and decision-making. According to Gibson (2006), the overall aim of the approach is to make 'net' or 'positive' contributions to sustainability based on explicit trade-offs among decision criteria, rather than merely identifying acceptable undertakings based on the avoidance of unwanted environmental effects. Still other authors (e.g., Scrase and Sheate 2002; Kidd and Fischer 2005; Morrison-Saunders and Fischer 2006) have criticized such 'integrated' approaches to assessment for encouraging the comparison of environmental and socioeconomic considerations, hence legitimizing environmental trade-offs for economic gain, and the social and environmental benefits thought to flow from such gain.

Here we rely on the definition of EIA's central task offered by Greig and Duinker (2011): "to prevent unacceptable impacts to the environment to help society achieve a sustainable pattern of development". In our view, the ultimate aim of making human developments more sustainable demands the protection of important ecological (and social) values, embodied in EIA by the concept of the VEC.

6.7 Climate change

The notion that human activities may be driving changes in Earth's global weather patterns began to gain traction in the late 1980s (e.g., Ramanathan et al. 1985; Ramanathan 1988; Mitchell 1989). Some of the first scientific research exploring the potential ecological consequences of global climate change was conducted at this time (e.g., Clark 1988; Pastor and Post 1988; Smith and Tirpak 1989). It was not until the early 1990s – at the United Nations Earth Summit in Rio – that governments around the world began to take serious action on global climate change. Here foundations were laid for an international agreement – the United Nations Framework Convention on Climate Change – aimed at the "stabilization of greenhouse gas concentrations in the atmosphere at a level that would prevent dangerous anthropogenic interference with the climate system" (UNFCCC 1992).

As part of the Rio climate convention, the UNFCCC (1992) identified two general kinds of implementable measures for dealing with global climate change and its effects: mitigation and adaptation. Mitigation measures would be aimed at minimizing the impact of human activities on global weather patterns by reducing greenhouse gas emissions. Adaptation measures, on the other hand, would be aimed at reducing social and ecological vulnerabilities to a changing climate through careful foresight and planning. The original UNFCCC (1992) agreement was later expanded by the Kyoto Protocol (UNFCCC 1998), which would explicitly commit state parties to reducing their greenhouse gas emissions in ways that reflect national differences (e.g., wealth, current emissions).

Though the EIA community was quick to endorse the notion of EIA as tool for addressing global climate change in the 1990s (e.g., Robinson 1992; Sadler 1996; CEQ 1997), practical guidance on the subject has only recently begun to emerge (e.g., CCCEAC 2003; CEQ 2010; Agrawala et al. 2010; Byer et al. 2012; Murphy and Gillam 2013). At the same time, reviews of practice (e.g., Sok et al. 2011; Slotterback 2011; Wende et al. 2012; Watkins and Durning 2012; Chang and Wu 2013; Kamau and Mwaura 2013; Ohsawa and Duinker 2014; Jiricka et al. 2016; Enríquez-de-Salamanca et al. 2016) have highlighted a variety of challenges surrounding the incorporation of climate change considerations into EIA, particularly at the project level.

Most scholarly reviews and guidance materials now acknowledge that it is impossible to predict how much global temperature would increase due to greenhouse gas emissions from a single project. Canada's federal guidelines for incorporating climate change into EIA (CCCEAC 2003) state that "unlike most project-related environmental effects, the contribution of an individual project to climate change cannot be measured". Though it may be relatively simple to quantify the greenhouse gas emissions associated with a particular development using techniques like life cycle analysis (e.g., Odeh and Cockerill 2008) or carbon budget modelling (e.g., Kurz et al. 2009), the task of evaluating the significance of such emissions continues to present a considerable challenge to practitioners. Many scholars (e.g., Christopher

2008; Slotterback 2011; Wende et al. 2012; Ohsawa and Duinker 2014) now agree that the gap between global or national emissions reduction targets and the emissions of local projects makes it very difficult to attribute any meaningful level of significance to such emissions. What is needed, they argue, is a kind of tiered EIA regime, whereby regional- or strategic-level EIAs provide a sort of backdrop for project-level EIAs. In this way, the significance of project-level emissions may be more meaningfully evaluated in terms of local, regional, national, and global emissions reductions targets, and then mitigated accordingly (i.e., best available technology, compensatory measures).

With respect to current EIA practice, several scholarly reviews (e.g., Kamau and Mwaura 2013; Ohsawa and Duinker 2014; Jiricka et al. 2016) have observed an overwhelming focus on climate change mitigation. Jiricka et al. (2016) write that "To date, greater attention has been paid to climate change mitigation (reduction of greenhouse gases) than to the adaptation to climate change effects". In terms of adaptation, many authors (e.g., Slotterback 2011; Jiricka et al. 2016) have pointed out that a major challenge surrounding the incorporation of climate change into environmental impact predictions has been a high degree of uncertainty surrounding the precise local and regional consequences of a changing global climate. To improve explorations of the future in cases of considerable uncertainty (e.g., climate change), Duinker and Greig (2007) have proposed greater use of scenarios and scenario analysis in formal EIA practice. In the words of Duinker and Greig (2007): "We need to include in the set of an EIA's scenarios a comprehensive range of potential future developments, and all the key driving forces, such as climate change and human demographics, that can measurably affect the VECs".

In our view, project-level EIA has the potential to become an important tool for addressing issues of global climate change, both in terms of mitigation and adaptation. We therefore expect to see climate change considerations increasingly incorporated into EIAs conducted at the project level in the near future. We also anticipate that if formal project EIA is to contribute effectively to climate change mitigation efforts, it will need to be nested within broader EIA regimes operating at regional or strategic levels. In this way, the significance of project-level emissions may be more meaningfully judged in the context of regional-scale emissions inventories and targets for reduction. With respect to climate change adaptation, we observe that project-level EIA is already well suited to such a task, since EIA processes were originally intended to identify, evaluate, and select development alternatives based on a variety of environmental considerations. We point out that failing to consider the effects of climate change in project EIAs may seriously invalidate environmental impact predictions. Despite a considerable degree of uncertainty surrounding the local environmental consequences of global climate change, we urge all practitioners of EIA to acknowledge, characterize, and explore the uncertainties associated with climate change using innovative techniques like scenario analysis.

6.8 Ecosystem services

The notion of ecosystem services was popularized in the early 2000s by the United Nations Millennium Ecosystem Assessment, which defined them as "the benefits people obtain from ecosystems" (MEA 2003). The report went on to differentiate four broad categories of ecosystem services, namely: "provisioning services such as food and water; regulating services such as flood and disease control; cultural services such as spiritual, recreational, and cultural benefits; and supporting services, such as nutrient cycling, that maintain the conditions for life on Earth". The notion of ecosystem services was intended to provide a conceptual framework for understanding critical relationships linking ecosystems, human well-being, and environmental decision-making. According to the report, "This enables a decision process to determine which service or set of services is valued most highly and how to develop approaches to maintain services by managing the system sustainably" (MEA 2003).

Since then, several authors (e.g., Geneletti 2013; Helming et al. 2013; Honrado et al. 2013; Partidario and Gomes 2013; Rega and Spaziante 2013) have explored the possibility of using the ecosystem services concept to frame both strategic and project-level EIAs. While some authors (e.g., Karjalainen et al. 2013; Rosa and Sánchez 2016) have praised the approach for its emphasis on linking ecosystemic properties to human benefits and values, others (e.g., Baker et al. 2013) have argued that "the use of ecosystem services language may not resonate well with all stakeholders". Most authors now agree that there has been limited experience applying the concept of ecosystem services in formal EIA practice.

In addition to highlighting differences of opinion on the subject, Baker et al. (2013) delineate two broad approaches to incorporating ecosystem services in EIA: comprehensive and philosophical. According to Baker et al. (2013), "The former is marked by the more quantitative approach to ecosystem services – this may include a systematic identification of ecosystem service supply and demand across an area and may extend to the valuation of ecosystem services". In contrast, Baker et al. (2013) write that "The ecosystem service philosophy is more about the use of ecosystem services as a heuristic or as a framing for the environment". With respect to the philosophical approach, Baker et al. (2013) conclude that "it is a less significant departure from existing practice and relies on a changing of language and elements of the approach".

Here we argue that EIA's central task – protecting the sustainability of VECs – is entirely compatible with an ecosystem services perspective, given the chosen ecosystem services are of interest to some or all EIA participants.

References

Agrawala, S., Matus Kramer, A., Prudent-Richard, G., and Sainsbury, M. 2010. *Incorporating Climate Change Impacts and Adaptation in Environmental Impact Assessments: Opportunities and Challenges.* OECD Environmental Working Paper No. 24. Paris: OECD Publishing. 37 p.

Akçakaya, H.R. 2001. Linking population-level risk assessment with landscape and habitat models. *Science of the Total Environment* 274(1–3): 283–291.

Andersen, T., Carstensen, J., Hernández-García, E., and Duarte, C.M. 2009. Ecological thresholds and regime shifts: Approaches to identification. *Trends in Ecology & Evolution* 24(1): 49–57.

Arrow, K., Bolin, B., Costanza, R., Dasgupta, P., Folke, C., Holling, C.S., Jansson, B., Levin, S., Mäler, K., Perrings, C., et al. 1995. Economic growth, carrying capacity, and the environment. *Science* 268(5210): 520–521.

Baker, J., Sheate, W.R., Phillips, P., and Eales, R. 2013. Ecosystem services in environmental assessment – help or hindrance? *Environmental Impact Assessment Review* 40: 3–13.

Beanlands, G.E., and Duinker, P.N. 1983. *An Ecological Framework for Environmental Impact Assessment in Canada.* Halifax: Institute for Resource and Environmental Studies, Dalhousie University. 132 p.

Bedford, B.L., and Preston, E.M. 1988. Developing the scientific basis for assessing cumulative effects of wetland loss and degradation on landscape functions: Status, perspectives, and prospects. *Environmental Management* 12(5): 751–771.

Benson, M.H., and Garmestani, A.S. 2011. Embracing panarchy, building resilience and integrating adaptive management through a rebirth of the National Environmental Policy Act. *Journal of Environmental Management* 92(5): 1420–1427.

Bestelmeyer, B.T. 2006. Threshold concepts and their use in rangeland management and restoration: The good, the bad, and the insidious. *Restoration Ecology* 14(3): 325–329.

Bond, A.J., Morrison-Saunders, A., Gunn, J.A.E., Pope, J., and Retief, F. 2015. Managing uncertainty, ambiguity and ignorance in impact assessment by embedding evolutionary resilience, participatory modeling and adaptive management. *Journal of Environmental Management* 151: 97–104.

Brenden, T.O., Wang, L., and Su, Z. 2008. Quantitative identification of disturbance thresholds in support of aquatic resource management. *Environmental Management* 42(5): 821–832.

Burgess, R.L., and Sharpe, D.M., editors. 1981. *Forest Island Dynamics in Man-Dominated Landscapes.* Now York: Springer-Verlag. 310 p.

Byer, P., Cestti, R., Croal, P., Fisher, W., Hazell, S., and Kolhoff, A. 2012. *Climate Change in Impact Assessment: International Best Practice Principles.*

56   *Emerging concepts for science in EIA*

Special Publication Series No. 8. Fargo: International Association for Impact Assessment. 4 p.

Cairns, J. 1977. Aquatic ecosystem assimilative capacity. *Fisheries* 2(2): 5–7.

Canter, L.W., and Canty, G.A. 1993. Impact significance determination – basic considerations and a sequenced approach. *Environment Impact Assessment Review* 13(5): 275–297.

CCCEAC (Committee on Climate Change and Environmental Assessment in Canada). 2003. *Incorporating Climate Change Considerations in Environmental Assessment: General Guidance for Practitioners* [Internet]; [cited 2016 December 12]. Available from: http://ceaa.gc.ca/Content/A/4/1/%20A41F45C5-1A79-44FA-9091-D251EEE18322/Incorporating_Climate_Change_Considerations_in_Environmental_Assessment.pdf

CEAA (Canadian Environmental Assessment Agency). 1996. *Guide on Biodiversity and Environmental Assessment.* Hull: Canadian Environmental Assessment Agency. 15 p.

CEQ (Council on Environmental Quality). 1993. *Incorporating Biodiversity Considerations into Environmental Impact Analysis under the National Environmental Policy Act.* Washington, DC: Council on Environmental Quality. 29 p.

CEQ (Council on Environmental Quality). 1997. *Guidance Regarding Consideration of Global Climatic Change in Environmental Documents Prepared Pursuant to the National Environmental Policy Act* [Internet]; [cited 2016 December 12]. Available from: www.boem.gov/uploadedFiles/BOEM/Environmental_Steward ship/Environmental_Assessment/ceqmemo.pdf

CEQ (Council on Environmental Quality). 2010. *Draft NEPA Guidance on Consideration of the Effects of Climate Change and Greenhouse Gas Emissions* [Internet]; [cited 2016 December 12]. Available from: www.whitehouse.gov/sites/default/files/microsites/ceq/20100218-nepa-consideration-effects-ghg-draft-guidance.pdf

Chang, I., and Wu, J. 2013. Integration of climate change considerations into environmental impact assessment – implementation, problems and recommendations for China. *Frontiers of Environmental Science & Engineering* 7(4): 598–607.

Christopher, C.W. 2008. Success by a thousand cuts: The use of environmental impact assessment in addressing climate change. *Vermont Journal of Environmental Law* 9: 549–613.

Clark, J.S. 1988. Effect of climate change on fire regimes in northwestern Minnesota. *Nature* 334(6179): 233–235.

Conover, S., Strong, K., Hickey, T., and Sander, F. 1985. An evolving framework for environmental impact analysis. I: Methods. *Journal of Environmental Management* 21: 343–358.

Costanza, R., Sklar, F.H., White, M.L., and Day, Jr J.W. 1988. A dynamic spatial simulation model of land loss and marsh succession in coastal Louisiana. In: *Wetland Modeling: Developments in Environmental Modeling 12.* Mitsch, W.J., Straskraba, M., and Jørgensen, S.E., editors. Amsterdam: Elsevier. 99–114 p.

Diaz, M., Illera, J.C., and Hedo, D. 2001. Strategic environmental assessment of plans and programs: A methodology for estimating effects on biodiversity. *Environmental Management* 28(2): 267–279.

Duinker, P.N., Burbidge, E.L., Boardley, S.R., and Greig, L.A. 2013. Scientific dimensions of cumulative effects assessment: Toward improvements in guidance for practice. *Environmental Reviews* 21(1): 40–52.

Duinker, P.N., and Greig, L.A. 2006. The impotence of cumulative effects assessment in Canada: ailments and ideas for redeployment. *Environmental Management* 37(2): 153–161.

Duinker, P.N., and Greig, L.A. 2007. Scenario analysis in environmental impact assessment: improving explorations of the future. *Environmental Impact Assessment Review* 27(3): 206–219.

Ehrlich, A., and Ross, W. 2015. The significance spectrum and EIA significance determinations. *Impact Assessment and Project Appraisal* 33(2): 87–97.

Elton, C.S. 1958. *The Ecology of Invasions by Animals and Plants*. London: Methuen. 181 p.

Enríquez-de-Salamanca, Á., Martín-Aranda, R.M., and Díaz-Sierra, R. 2016. Consideration of climate change on environmental impact assessment in Spain. *Environmental Impact Assessment Review* 57: 31–39.

Ficetola, G.F., and Denoël, M. 2009. Ecological thresholds: An assessment of methods to identify abrupt changes in species – habitat relationships. *Ecography* 32(6): 1075–1084.

Folke, C., Carpenter, S., Walker, B., Scheffer, M., Elmqvist, T., Gunderson, L., and Holling, C.S. 2004. Regime shifts, resilience, and biodiversity in ecosystem management. *Annual Review of Ecology, Evolution, and Systematics* 35(1): 557–581.

Folke, C., Carpenter, S., Elmqvist, T., Gunderson, L., Holling, C.S., and Walker, B. 2002. Resilience and sustainable development: Building adaptive capacity in a world of transformations. *Ambio* 31(5): 437–440.

Forman, R.T., and Godron, M. 1981. Patches and structural components for a landscape ecology. *Bioscience* 31(10): 733–740.

Forman, R.T., and Godron, M. 1986. *Landscape Ecology*. New York: John Wiley & Sons. 619 p.

Franklin, J.F. 1988. Structural and functional diversity in temperate forests. In: *Biodiversity*. Wilson, E.O. and Peter, F.M., editors. Washington, DC: National Academy Press. 166–175 p.

Gardner, J.E. 1989. Decision making for sustainable development: Selected approaches to environmental assessment and management. *Environmental Impact Assessment Review* 9(4): 337–366.

Gardner, J.E. 1990. The elephant and the nine blind men: An initial review of environmental assessment and related processes in support of sustainable development. In: *Sustainable Development and Environmental Assessment: Perspectives on Planning for a Common Future*. Jacobs, P. and Sadler, B., editors. Hull: Canadian Environmental Assessment Research Council. 35–66 p.

Gause, G.F., and Smaragdova, N.P., and Witt, A.A. 1936. Further studies of interaction between predators and prey. *Journal of Animal Ecology* 5(1): 1–18.

Geneletti, D. 2003. Biodiversity impact assessment of roads: An approach based on ecosystem rarity. *Environmental Impact Assessment Review* 23(3): 343–365.

Geneletti, D. 2013. Assessing the impact of alternative land-use zoning policies on future ecosystem services. *Environmental Impact Assessment Review* 40: 25–35.

Gibson, R.B. 2006. Sustainability assessment: Basic components of a practical approach. *Impact Assessment and Project Appraisal* 24(3): 170–182.

Gibson, R.B., Hassan, S., Holtz, S., Tansey, J., and Whitelaw, G. 2005. *Sustainability Assessment: Criteria and Processes*. London: Earthscan. 254 p.

Gontier, M., Balfors, B., and Mörtberg, U. 2006. Biodiversity in environmental assessment – current practice and tools for prediction. *Environmental Impact Assessment Review*. 26(3): 268–286.

Greig, L.A., and Duinker, P.N. 2011. A proposal for further strengthening science in environmental impact assessment in Canada. *Impact Assessment Project Appraisal* 29(2): 159–165.

Groffman, P.M., Baron, J.S., Blett, T., Gold, A.J., Goodman, I., Gunderson, L.H., Levinson, B.M., Palmer, M.A., Paerl, H.W., and Peterson, G.D., et al. 2006. Ecological thresholds: the key to successful environmental management or an important concept with no practical application? *Ecosystems* 9(1): 1–13.

Gunderson, L. 1999. Resilience, flexibility and adaptive management – antidotes for spurious certitude? *Ecology and Society* 3(1): 7.

Gunderson, L.H., and Holling, C.S., editors. 2001. *Panarchy: Understanding Transformations in Human and Natural Systems*. Washington, DC: Island press. 507 p.

Gustafson, E.J., Shifley, S.R., Mladenoff, D.J., Nimerfro, K.K., and He, H.S. 2000. Spatial simulation of forest succession and timber harvesting using LANDIS. *Canadian Journal of Forest Research* 30(1): 32–43.

Hansson, L. 1977. Landscape ecology and stability of populations. *Landscape Plan* 4: 85–93.

Haug, P., Burwell, R., Stein, A., and Bandurski, B. 1984. Determining the significance of environmental issues under the National Environmental Policy Act. *Journal of Environmental Management* 18: 15–24.

Hecky, R.E., Newbury, R.W., Bodaly, R.A., Patalas, K., and Rosenberg, D.M. 1984. Environmental impact prediction and assessment: The Southern Indian Lake experience. *Canadian Journal of Fisheries and Aquatic Sciences* 41(4): 720–732.

Helming, K., Diehl, K., Geneletti, D., and Wiggering, H. 2013. Mainstreaming ecosystem services in European policy impact assessment. *Environmental Impact Assessment Review* 40: 82–87.

Holling, C.S. 1966. The functional response of invertebrate predators to prey density. *Memoirs of the Entomological Society of Canada* 98: 5–86.

Holling, C.S. 1973. Resilience and stability of ecological systems. *Annual Review Ecology and Systematics* 4: 1–23.

Holling, C.S. 1986. The resilience of terrestrial ecosystems: Local surprise and global change. In: *Sustainable Development of the Biosphere*. Clark, W.C. and Munn, R.E., editors. Cambridge: Cambridge University Press. 292–317 p.

Holling, C.S. 2001. Understanding the complexity of economic, ecological, and social systems. *Ecosystems* 4(5): 390–405.

Holling, C.S., editor. 1978. *Adaptive Environmental Assessment and Management*. Toronto: John Wiley & Sons. 377 p.

Honrado, J.P., Vieira, C., Soares, C., Monteiro, M.B., Marcos, B., Pereira, H.M., and Partidário, M.R. 2013. Can we infer about ecosystem services from EIA and SEA practice? A framework for analysis and examples from Portugal. *Environmental Impact Assessment Review* 40: 14–24.

Houle, M., Fortin, D., Dussault, C., Courtois, R., and Ouellet, J. 2010. Cumulative effects of forestry on habitat use by gray wolf (*Canis lupus*) in the boreal forest. *Landscape Ecology* 25(3): 419–433.

Hunter, M.L., Jacobson, G.L., and Webb, T. 1988. Paleoecology and the coarse-filter approach to maintaining biological diversity. *Conservation Biology* 2(4): 375–385.

Hurlbert, S.H. 1971. The non-concept of species diversity: A critique and alternative parameters. *Ecology* 52(4): 577–586.

Huston, M. 1979. A general hypothesis of species diversity. *The American Naturalist* 113(1): 81–101.

Jacobs, P., and Sadler, B., editors. 1990. *Sustainable Development and Environmental Assessment: Perspectives on Planning for a Common Future.* Hull: Canadian Environmental Assessment Research Council. 182 p.

Jiricka, A., Formayer, H., Schmidt, A., Völler, S., Leitner, M., Fischer, T.B., and Wachter, T.F. 2016. Consideration of climate change impacts and adaptation in EIA practice – perspectives of actors in Austria and Germany. *Environmental Impact Assessment Review* 57: 78–88.

Johnson, C.J. 2013. Identifying ecological thresholds for regulating human activity: Effective conservation or wishful thinking? *Biological Conservation* 168: 57–65.

Kamau, J.W., and Mwaura, F. 2013. Climate change adaptation and EIA studies in Kenya. *International Journal of Climate Change Strategies and Management* 5(2): 152–165.

Karjalainen, T.P., Marttunen, M., Sarkki, S., and Rytkönen, A. 2013. Integrating ecosystem services into environmental impact assessment: An analytic – deliberative approach. *Environmental Impact Assessment Review* 40: 54–64.

Karr, J.R. 1981. Assessment of biotic integrity using fish communities. *Fisheries* 6(6): 21–27.

Kay, J.J., Regier, H.A., Boyle, M., and Francis, G. 1999. An ecosystem approach for sustainability: Addressing the challenge of complexity. *Futures* 31(7): 721–742.

Kessell, S.R., Good, R.B., and Hopkins, A.J.M. 1984. Implementation of two new resource management information systems in Australia. *Environmental Management* 8(3): 251–269.

Khera, N., and Kumar, A. 2010. Inclusion of biodiversity in environmental impact assessments (EIA): A case study of selected EIA reports in India. *Impact Assessment and Project Appraisal* 28(3): 189–200.

Kidd, S., and Fischer, T.B. 2007. Towards sustainability: Is integrated appraisal a step in the right direction? *Environmental and Planning C* 25(2): 233–249.

Kjellerup, U. 1999. Significance determination: A rational reconstruction of decisions. *Environmental Impact Assessment Review* 19(1): 3–19.

Kurz, W.A., Dymond, C.C., White, T.M., Stinson, G., Shaw, C.H., Rampley, G.J., Smyth, C., Simpson, B.N., Neilson, E.T., Trofymow, J.A., et al. 2009. CBM-CFS3: A model of carbon-dynamics in forestry and land-use change implementing IPCC standards. *Ecological Modeling* 220(4): 480–504.

60 *Emerging concepts for science in EIA*

Landres, P.B. 1983. Use of the guild concept in environmental impact assessment. *Environmental Management* 7(5): 393–397.

Larkin, P.A., Raleigh, R.F., and Wilimovsky, N.J. 1964. Some alternative premises for constructing theoretical reproduction curves. *Journal of Fisheries Research Board of Canada* 21(3): 477–484.

Larson, M.A., Thompson III, F.R., Millspaugh, J.J., Dijak, W.D., Shifley, S.R. 2004. Linking population viability, habitat suitability, and landscape simulation models for conservation planning. *Ecological Modeling* 180(1): 103–118.

Lawrence, D.P. 1997. Integrating sustainability and environmental impact assessment. *Environmental Management* 21(1): 23–42.

Lee, L.C., and Gosselink, J.G. 1988. Cumulative impacts on wetlands: Linking scientific assessments and regulatory alternatives. *Environmental Management* 12(5): 591–602.

Levin, S.A., and Paine, R.T. 1974. Disturbance, patch formation, and community structure. *Proceedings of the National Academy of Sciences of the USA* 71(7): 2744–2747.

Lugo, A.E. 1988. Estimating reductions in the diversity of tropical forest species. In: *Biodiversity*. Wilson, E.O. and Peter, F.M., editors. Washington, DC: National Academy Press. 58–70 p.

MacArthur, R. 1955. Fluctuations of animal populations and a measure of community stability. *Ecology* 36(3): 533–536.

Mandelik, Y., Dayan, T., and Feitelson, E. 2005. Planning for biodiversity: The role of ecological impact assessment. *Conservation Biology* 19(4): 1254–1261.

May, R.M. 1973. *Stability and Complexity in Model Ecosystems*. Princeton: Princeton University Press. 235 p.

MEA (Millennium Ecosystem Assessment). 2003. *Ecosystems and Human Wellbeing: A Framework for Assessment*. Washington, DC: Island Press. 245 p.

Mitchell, J.F.B. 1989. The "greenhouse" effect and climate change. *Review of Geophysics* 27(1): 115–139.

Mladenoff, D.J., and He, H.S. 1999. Design, behavior and application of LANDIS, an object-oriented model of forest landscape disturbance and succession. In: *Spatial Modeling of Forest Landscape Change: Approaches and Applications*. Mladenoff, D.J. and Baker, W.L., editors. Cambridge: Cambridge University Press. 125–162 p.

Morrison-Saunders, A., and Fischer, T.B. 2006. What is wrong with EIA and SEA anyway? A sceptic's perspective on sustainability assessment. *Journal of Environmental Assessment Policy and Management* 8(1): 19–39.

Mörtberg, U.M., Balfors, B., and Knol, W.C. 2007. Landscape ecological assessment: A tool for integrating biodiversity issues in strategic environmental assessment and planning. *Journal of Environmental Management* 82(4): 457–470.

Murphy, M.C., and Gillam, K.M. 2013. *Greenhouse Gases and Climate in Environmental Impact Assessment – Practical Guidance*. IAIA13 Conference Proceedings [Internet]; [cited 2016 December 12]. Available from: http://conferences.iaia. org/2013/pdf/Final%20papers%20review%20process%2013/Greenhouse%20 Gases%20and%20Climate%20in%20Environmental%20Impact%20Assess ment%20%E2%80%93%20Practical%20Guidance.pdf

Myers, N. 1988. Tropical forests and their species: going, going . . .? In: *Biodiversity*. Wilson, E.O. and Peter, F.M., editors. Washington, DC: National Academy Press. 28–35 p.

Naveh, Z. 1982. Landscape ecology as an emerging branch of human ecosystem science. *Advances in Ecological Research* 12: 189–237.

Nelson, J.G., and Serafin, R. 1991. *Biodiversity and Environmental Assessment.* Waterloo: International Union for the Conservation of Nature. 13 p.

Nelson, J.G., and Serafin, R. 1992. Assessing biodiversity: A human ecological approach. *Ambio* 21(3): 212–218.

Noss, R.F. 1987. From plant communities to landscapes in conservation inventories: A look at the nature conservancy (USA). *Biological Conservation* 41(1): 11–37.

Odeh, N.A., and Cockerill, T.T. 2008. Life cycle GHG assessment of fossil fuel power plants with carbon capture and storage. *Energy Policy* 36(1): 367–380.

Ohsawa, T., and Duinker, P. 2014. Climate-change mitigation in Canadian environmental impact assessments. *Impact Assessment Project Appraisal* 32(3): 222–233.

Pahl-Wostl, C. 2007. The implications of complexity for integrated resources management. *Environmental Modeling & Software* 22(5): 561–569.

Partidario, M.R., and Gomes, R.C. 2013. Ecosystem services inclusive strategic environmental assessment. *Environmental Impact Assessment Review* 40: 36–46.

Pastor, J., and Post, W.M. 1988. Response of northern forests to CO_2-induced climate change. *Nature* 334(6177): 55–58.

Pearlstine, L., McKellar, H., and Kitchens, W. 1985. Modeling the impacts of a river diversion on bottomland forest communities in the Santee River floodplain, South Carolina. *Ecological Modeling* 29(1): 283–302.

Peet, R.K. 1974. The measurement of species diversity. *Annual Review of Ecology Systematics* 5: 285–307.

Peterson, G., Allen, C.R., and Holling, C.S. 1998. Ecological resilience, biodiversity, and scale. *Ecosystems* 1(1): 6–18.

Pope, J., Annandale, D., and Morrison-Saunders, A. 2004. Conceptualising sustainability assessment. *Environmental Impact Assessment Review* 24(6): 595–616.

Puettmann, K.J., Messier, C.C., and Coates, K.D. 2009. *A Critique of Silviculture: Managing for Complexity.* Washington, DC: Island Press. 189 p.

Ramanathan, V. 1988. The greenhouse theory of climate change: A test by an inadvertent global experiment. *Science* 240(4850): 293–299.

Ramanathan, V., Cicerone, R.J., Singh, H.B., and Kiehl, J.T. 1985. Trace gas trends and their potential role in climate change. *Journal of Geophysical Research-Atmospheres* 90(D3): 5547–5566.

Ray, C. 1988. Ecological diversity in coastal zones and oceans. In: *Biodiversity*. Wilson, E.O. and Peter, F.M., editors. Washington, DC: National Academy Press. 36–50 p.

Rees, W.E. 1988. A role for environmental assessment in achieving sustainable development. *Environmental Impact Assessment Review* 8(4): 273–291.

Rega, C., and Spaziante, A. 2013. Linking ecosystem services to agri-environmental schemes through SEA: A case study from Northern Italy. *Environmental Impact Assessment Review* 40: 47–53.

Robinson, N.A. 1992. International trends in environmental impact assessment. *Boston College Environmental Affairs Law Review* 19: 591–622.

Rosa, J.C.S., and Sánchez, L.E. 2016. Advances and challenges of incorporating ecosystem services into impact assessment. *Journal of Environmental Management* 180: 485–492.

Sadler, B. 1996. *Environmental Assessment in a Changing World: Evaluating Practice to Improve Performance*. Final Report of the International Study of the Effectiveness of Environmental Assessment. Hull: Canadian Environmental Assessment Agency. 248 p.

Sassaman, R.W. 1981. Threshold of concern: A technique for evaluating environmental impacts and amenity values. *Journal of Forestry* 79(2): 84–86.

Scheffer, M., and Carpenter, S.R. 2003. Catastrophic regime shifts in ecosystems: Linking theory to observation. *Trends in Ecology & Evolution* 18(12): 648–656.

Scheller, R.M., Domingo, J.B., Sturtevant, B.R., Williams, J.S., Rudy, A., Gustafson, E.J., and Mladenoff, D.J. 2007. Design, development, and application of LANDIS-II, a spatial landscape simulation model with flexible temporal and spatial resolution. *Ecological Modeling* 201(3–4): 409–419.

Scheller, R.M., and Mladenoff, D.J. 2004. A forest growth and biomass module for a landscape simulation model, LANDIS: design, validation, and application. *Ecological Modeling* 180(1): 211–229.

Schmidt, M., Glasson, J., Emmelin, L., and Helbron, H., editors. 2008. *Standards and Thresholds for Impact Assessment*. Berlin: Springer Berlin Heidelberg. 493 p.

Scrase, J.I., and Sheate, W.R. 2002. Integration and integrated approaches to assessment: What do they mean for the environment? *Journal of Environmental Political Plan* 4(4): 275–294.

Severinghaus, W.D. 1981. Guild theory development as a mechanism for assessing environmental impact. *Environmental Management* 5(3): 187–190.

Shugart, H.H., and West, D.C. 1981. Long-term dynamics of forest ecosystems. *American Scientist* 69(6): 647–652.

Sinclair, A.J., Diduck, A., and Fitzpatrick, P. 2008. Conceptualizing learning for sustainability through environmental assessment: Critical reflections on 15 years of research. *Environmental Impact Assessment Review* 28(7): 415–428.

Sklar, F.H., Costanza, R., and Day, J.W. 1985. Dynamic spatial simulation modeling of coastal wetland habitat succession. *Ecological Modeling* 29(1): 261–281.

Slootweg, R., and Jones, M. 2011. Resilience thinking improves SEA: A discussion paper. *Impact Assessment and Project Appraisal* 29(4): 263–276.

Slootweg, R., and Kolhoff, A. 2003. A generic approach to integrate biodiversity considerations in screening and scoping for EIA. *Environmental Impact Assessment Review* 23(6): 657–681.

Slootweg, R., Rajvanshi, A., Mathur, V.B., and Kolhoff, A, editors. 2010. *Biodiversity in Environmental Assessment: Enhancing Ecosystem Services for Human Well-Being*. Cambridge: Cambridge University Press. 437 p.

Slotterback, C.S. 2011. Addressing climate change in state and local environmental impact analysis. *Journal of Environmental Planning and management* 54(6): 749–767.

Smith, J.B., and Tirpak, D.A., editors. 1989. *The Potential Effects of Global Climate Change on the United States*. EPA-230-05-89-050. Washington, DC: US Environmental Protection Agency. 413 p.

Söderman, T. 2006. Treatment of biodiversity issues in impact assessment of electricity power transmission lines: A Finnish case review. *Environmental Impact Assessment Review* 26(4): 319–338.

Sok, V., Boruff, B.J., and Morrison-Saunders, A. 2011. Addressing climate change through environmental impact assessment: International perspectives from a survey of IAIA members. *Impact Assessment and Project Appraisal* 29(4): 317–325.

Sorensen, T., McLoughlin, P.D., Hervieux, D., Dzus, E., Nolan, J., Wynes, B., and Boutin, S. 2008. Determining sustainable levels of cumulative effects for boreal caribou. *Journal of Wildlife Management* 72(4): 900–905.

Troll, C. 1971. Landscape ecology (geoecology) and biogeocenology – a terminological study. *Geoforum* 2(4): 43–46.

Turner, M.G. 1989. Landscape ecology: The effect of pattern on process. *Annual Review of Ecology, Evolution, and Systematics* 20: 171–197.

UNCED (United Nations Conference on Environment and Development). 1992. *Agenda 21: The United Nations Programme of Action From Rio*. New York: United Nations. 300 p.

UNFCCC (United Nations Framework Convention on Climate Change). 1992. *United Nations Framework Convention on Climate Change* [Internet]; [cited 2016 December 12]. Available from: https://unfccc.int/resource/docs/convkp/con veng.pdf

UNFCCC (United Nations Framework Convention on Climate Change). 1998. *Kyoto Protocol to the United Nations Framework Convention on Climate Change* [Internet]; [cited 2016 December 12]. Available from: https://unfccc.int/resource/docs/convkp/kpeng.pdf

Urban, D.L., O'Neill, R.V., and Shugart, H.H. 1987. Landscape ecology. *Bioscience* 37(2): 119–127.

Wale, E., and Yalew, A. 2010. On biodiversity impact assessment: The rationale, conceptual challenges and implications for future EIA. *Impact Assessment and Project Appraisal* 28(1): 3–13.

Ware, D.M. 1972. Predation by rainbow trout (*Salmo gairdneri*): the influence of hunger, prey density, and prey size. *Journal of Fisheries Research Board of Canada* 29(8): 1193–1201.

Watkins, J., and Durning, B. 2012. Carbon definitions and typologies in environmental impact assessment: Greenhouse gas confusion? *Impact Assessment and Project Appraisal* 30(4): 296–301.

WCED (World Commission on Environment and Development). 1987. *Our Common Future*. Oxford: Oxford University Press. 383 p.

Weaver, P.M., and Rotmans, J. 2006. Integrated sustainability assessment: What is it, why do it and how? *International Journal of Innovation and Sustainable Development* 1(4): 284–303.

Wegner, A., Moore, S.A., and Bailey, J. 2005. Consideration of biodiversity in environmental impact assessment in Western Australia: Practitioner perceptions. *Environmental Impact Assessment Review* 25(2): 143–162.

Weller, M.W. 1988. Issues and approaches in assessing cumulative impacts on water bird habitat in wetlands. *Environmental Management* 12(5): 695–701.

Wende, W., Bond, A., Bobylev, N., and Stratmann, L. 2012. Climate change mitigation and adaptation in strategic environmental assessment. *Environmental Impact Assessment Review* 32(1): 88–93.

Whittaker, R.H. 1972. Evolution and measurement of species diversity. *Taxon* 21(2): 213–251.

Wiens, J.A. 1976. Population responses to patchy environments. *Annual Review of Ecology Systematics* 7: 81–120.

Wilson, E.O., and Peter, F.M., editors. 1988. *Biodiversity*. Washington, DC: National Academy Press. 521 p.

Winter, T.C. 1988. A conceptual framework for assessing cumulative impacts on the hydrology of non-tidal wetlands. *Environmental Management* 12(5): 605–620.

Wong, S.L., and Clark, B. 1976. Field determination of the critical nutrient concentrations for *Cladophora* in streams. *Journal of Fisheries Research Board of Canada* 33(1): 85–92.

7 Science in the EIA process

7.1 Scoping

7.1.1 Overview

In EIA processes, scoping refers to an early set of activities aimed at defining a stakeholder-relevant focus for assessments. Through scoping, professionals and stakeholders identify concerns surrounding proposed developments and agree on the most important environmental values at stake. As with the formal review stage of EIA processes, scoping implies the convergence of political, scientific, and administrative expectations and perspectives. From an administrative perspective, scoping produces a list of proposal-specific guidelines that direct the proponent in preparing an environmental impact statement (EIS) or similar EIA document. From a political perspective, scoping provides an early opportunity for all interested parties to influence the general design of assessments prior to their implementation and review. From a scientific perspective, scoping aims to focus assessments on a well-defined set of questions for which targeted research programs can be designed and implemented. Overall, the scoping process is aimed at fostering early agreement among professionals and stakeholders alike as to how an EIA will proceed and on what basis scientific contributions to that EIA will be evaluated during review.

Originally, formal EIA processes did not require an initial scoping exercise to focus assessments. The tendency of these early processes was to produce voluminous and excessively descriptive documentation, a phenomenon which was repeatedly observed in the literature of the 1970s (e.g., Andrews 1973; Carpenter 1976; Schindler 1976; Holling 1978; Ward 1978; Munn 1979). Indeed, early frameworks for conducting EIAs (e.g., Leopold et al. 1971; Dee et al. 1973) advocated comprehensive coverage of environmental entities, regardless of their relevance to development decisions.

Subsequent scientific guidance materials (e.g., Holling 1978; Ward 1978; Munn 1979) and regulatory provisions (e.g., CEQ 1978) called for more focused EIAs predicated on early and collaborative problem-structuring exercises. While Holling (1978) referred to such early and collaborative problem structuring as 'bounding', official EIA guidelines (e.g., CEQ 1980) would later to refer to it as 'scoping'. Such guidance sought to shift the initial point of contact between stakeholders and professionals away from adversarial reviews to focus on more creative aspects of assessment planning and design.

In the 1980s, scientific guidance materials for EIA (e.g., Beanlands and Duinker 1983; Beanlands 1988) thoughtfully elaborated on the principles of scoping outlined by administrators in the late 1970s. Most notably, Beanlands and Duinker (1983) coined the term 'VEC' to encourage those responsible for preparing EIAs to focus their investigations on environmental entities of interest to public and professional stakeholders. Beanlands and Duinker (1983) recognized that the immediate goal of EIA – to protect environmental values by influencing development decisions – would be more likely to succeed if scientific contributions were to focus explicitly on the issues of greatest concern to citizens. In addition to VEC-focused assessments, Beanlands and Duinker (1983) called for greater participation from the scientific research community during the early stages of assessment planning and design. In sum, Beanlands and Duinker (1983) outlined a framework for designing and implementing collaborative research programs aimed at providing useful and defensible insights into stakeholder-relevant issues.

The literature of the 1990s and 2000s has continued to emphasize the importance of good scoping in determining both the quality of EIAs and the overall acceptability of EIA-related development decisions (e.g., Wood 1995; Jones 1999; Treweek 1999). Still, scoping has been identified as an ongoing area of weakness in EIA practice from both technical and participatory perspectives (e.g., Wood 1995; Sadler 1996; Weston 2000; Ross et al. 2006; Snell and Cowell 2006; Wood et al. 2006). Despite widespread agreement on a basic definition for scoping, two competing approaches to scoping seem to have emerged in the literature. For some (e.g., Kennedy and Ross 1992; Ross et al. 2006), scoping functions as a sort of 'reversible funnel' for adding ('scoping in') and then removing ('scoping out') issues and VECs from a list of things to address during an assessment. For others (e.g., Mulvihill and Jacobs 1998; Mulvihill and Baker 2001; Mulvihill 2003), scoping functions as a creative and inclusive design process aimed at outlining a well-defined, stakeholder-relevant framework for investigation and evaluation. Indeed, scoping may be conducted under more formal and adversarial arrangements (e.g., guidelines hearings), or under more informal and collaborative arrangements (e.g., non-adversarial stakeholder meetings, scenario-building exercises).

More recently, Morrison-Saunders et al. (2014) have argued that current perceptions of the lack of efficiency and effectiveness in EIA practice can be attributed to the "proliferation of assessment types that has emerged over the past several decades". This, they argue, "is creating silos and confusion amongst regulators, stakeholders and even impact assessment practitioners". The authors go on to argue that: "the solution to the problem is to take an integrated approach which requires an emphasis in particular during the scoping stage". In response to the Morrison-Saunders et al. (2014) paper, Greig and Duinker (2014) have suggested that several other technical elements of EIA are also in need of repair and improvement before EIA can make the full contribution to sustainable development for which it was purportedly designed.

Here we provide a review of the literature surrounding the five major themes typically addressed through EIA scoping: (i) development alternatives, (ii) VECs, (iii) indicators, (iv) drivers, and (v) boundaries. We also provide a review of any literature that has helped to explain the role of science in addressing each of these five themes.

7.1.2 Alternatives

The consideration of technically and economically feasible development alternatives has been widely acknowledged as a foundational principle of EIA (e.g., Andrews 1973; CEQ 1978; Beanlands and Duinker 1983; Holling 1978; Wood 1995; Sadler 1996). It has also been recognized as an important opportunity to integrate EIA processes with the often separate enterprise of development planning and design (e.g., Brown and Hill 1995; McDonald and Brown 1995). Despite being a formally stated requirement in most EIA laws and policies, the identification and evaluation of alternatives has been widely recognized as a major area of weakness in EIA practice (e.g., Hill and Ortolano 1978; Wood 1995; Valve 1999; Steinemann 2001). While development alternatives are often described briefly at the beginning of assessments, they are rarely carried through to the impact prediction, significance evaluation, and review stages of the EIA process. Instead, alternative developments are typically dismissed early on, with little accompanying rationale to justify the selection of a preferred alternative.

The premise behind the consideration of alternatives in EIA is that, in general, the stated goal or purpose of a proposed development can often be achieved in different ways, each of which will have different environmental impacts. Depending on the nature of development proposals, four general types of alternatives may be defined at the outset of EIA processes:

(i)   the 'no-action' alternative;
(ii) alternative sites or locations for development;
(iii) alternative development designs (i.e., 'alternative means' of implementing the proposed development);
(iv) functionally distinct ways of achieving development goals (i.e., 'alternatives to' the proposed development).

From a scientific perspective, the consideration of alternatives in EIA allows for the identification of development configurations that will ensure the greatest protection of VECs. From a political perspective, the participatory design of alternatives enhances the likelihood of negotiating a mutually agreeable decision following EIA report submission and review. In short, the early identification of alternatives to be compared through impact prediction, evaluation, and review is an important element of a strong scoping process.

7.1.3 VECs

Given its ultimate purpose of securing sustainable development by protecting important environmental values (Sadler 1996), the EIA process must begin by explicitly identifying VECs based on stakeholder concerns (Beanlands and Duinker 1983). Broadly speaking, VECs are specified entities of the environment (e.g., particular species, specific ecosystems, water and air quality) that are valued by members of society but potentially compromised by proposed developments. For VECs to accurately embody the range of environmental values held by different stakeholder groups, they must be explicitly selected on the basis of public concern, professional concern, or both. According to Beanlands and Duinker (1983), while there is no sure way of anticipating the specific environmental concerns of stakeholders, human societies in general can be expected to place a high value on: (i) environmental components related to human health and safety; (ii) biotic species of major commercial, recreational, or aesthetic importance; (iii) rare or endangered species; and (iv) suitable habitat for such species. To be sure, scientific contributions to EIA will be far more useful to all participants if they focus explicitly on stakeholder-relevant VECs as well as the key ecosystem components and processes that sustain such VECs.

More recently, the notion of EIA as being focused on more traditional components of ecosystems has been expanded. With growing concerns over the consequences of global climate change and biodiversity loss, such issues have been increasingly considered in the context of formal EIA. While the issue of climate change is fairly specific, that of biodiversity loss implies the consideration of many biological entities operating at multiple levels of organization. The literature (e.g., Nelson and Serafin 1991; Gontier et al. 2006) indicates that the concept of biodiversity – as with that of the ecosystem – can be dissected into components that are both readily identifiable and valued by stakeholders and decision-makers alike. Gontier et al. (2006) refer to such entities as 'valued biodiversity components'.

7.1.4 Indicators

The term 'indicator' has been used in ecology and environmental planning circles for some time (e.g., Moore 1966) to mean any perceptible attribute of the biophysical environment that is to be measured repeatedly to detect environmental changes. Whereas some ecologists (e.g., Swank and Douglass 1975; Jackson et al. 1977) tended to focus on biogeochemical indicators (e.g., nutrient flux) that would reflect the aggregate condition of whole ecosystems, others (e.g., Cairns 1974; Bauerle et al. 1975; Ray and White 1976) tended to focus on biological indicators (e.g., species population, community structure) that would reflect the condition of specific ecosystem components (e.g., air and water quality).

Holling's (1978) treatise on AEAM outlined a collaborative approach to indicator selection, whereby scientists and decision-makers work together to identify measurable environmental attributes that reflect the interests and concerns of decision-makers. Holling (1978) wrote that indicators are "Measures of system behavior in terms of meaningful and perceptible attributes". According to Holling (1978), the bounding (i.e., scoping) stage of EIA processes was to be focused primarily on the identification of: (i) state variables and (ii) indicators. In Holling's (1978) view, state variables represent the key components needed to capture the behaviour of complex ecosystems, while indicators constitute the terms in which environmental effects (predicted or measured) are to be evaluated by decision-makers.

Decades later, Heink and Kowarik (2010) have observed confusion surrounding the term 'indicator' and its precise meaning in the context of environmental decision-making. The authors write that "The term 'indicator' is frequently used at the interface between science and policy. Although there is a great demand for clear definitions of technical terms in science and policy, the meaning of indicator is still ambiguous". Heink and Kowarik (2010) recommend distinguishing between indicators in five different ways:

(i) as ecosystem components (i.e., ecological units, structures, or processes) or measures (i.e., properties of a phenomenon, body, or substance to which a magnitude can be assigned);

(ii) as descriptive or normative (i.e., prescriptive) statements about ecological conditions;
(iii) as direct or indirect measures of an ecosystem component;
(iv) as simple indicators or complex (i.e., composite) indices;
(v) as measures of simple ecosystem components (e.g., wildlife populations) or more complex ones (e.g., biodiversity).

In our view, VECs (including both simple and complex ecological entities) embody the values and concerns of stakeholders and decision-makers.
Indicators, on the other hand, constitute the terms in which environmental
changes are predicted, communicated, and subsequently measured. Though
some authors have favoured the use of aggregate indices of environmental
impact (e.g., Dee et al. 1973; Antunes et al. 2001; Canter and Atkinson
2011), we note that the concept of indicators is much more basic. Indicators
are measurable attributes of the environment that describe the state or condition of VECs. While an individual VEC may have more than one indicator
attached to it, direct measurement of VEC condition is always preferred
over indirect measurement. In some instances, where direct measurement
of VEC condition is not feasible (e.g., rare species population), indirect
measurement of a related environmental attribute may suffice (e.g., suitable habitat). In such cases, the relationship between the selected indicator
and the VEC condition being measured ought to be clearly established and
periodically verified.

7.1.5 Boundaries

The importance of setting reasonable time and space boundaries for EIAs
has been widely acknowledged in the literature (e.g., Holling 1978; Duinker
and Baskerville 1986; Clark 1994; Noble 2000; João 2002; Geneletti 2006;
Karstens et al. 2007). It is agreed upon that the explicit delineation of study
boundaries in space and time provides a further level of specificity and
focus to the design and conduct of EIAs.

Beanlands and Duinker (1983) wrote that the delimitation of EIA studies
in space and time requires explicit consideration of multiple overlapping
human-imposed and naturally occurring boundaries. These include jurisdictional and study-length limitations imposed by administrative authorities,
the spatial extent and lifespan of proposed developments, and the temporal
and spatial scales at which ecosystems naturally operate. Within the limits imposed by administrative authorities, Beanlands and Duinker (1983)
argued that participants should begin by first considering an initial set of
spatial boundaries based on physical transport mechanisms (e.g., movement
of water and nutrients). Because physical boundaries do not always match

biological ones, Beanlands and Duinker (1983) argued that an initial consideration of physical dynamics in space should be followed by a consideration of biological dynamics in space (e.g., animal migration).

Beanlands and Duinker (1983) observed that reasonable study boundaries in time could be established based on a variety of temporal characteristics of ecosystems. According to Beanlands and Duinker (1983), "Such factors include: (i) the magnitude, periodicity and trends in the natural variation of the variables of interest, (ii) the time required for a biotic response to become evident, and (iii) the time required for a system or subsystem to recover from a perturbation to its pre-impact state". Beanlands and Duinker (1983) further likened such temporal ecological boundaries to the concept of resilience outlined by Holling (1973), as well as the concept of recoverability outlined by some other authors (e.g., Cairns 1980). Beanlands and Duinker (1983) concluded that although conceptually appealing, such theoretical notions had yet to find practical application in the detection of temporal ecosystem boundaries.

More recently, some authors (e.g., Clark 1994; João 2002; Geneletti 2006) have continued to explore the task of boundary-setting from a technical perspective, while others (e.g., Karstens et al. 2007) have begun to examine the issue from a participatory perspective. Since the Beanlands and Duinker (1983) report was published, however, very few guidance materials have emerged on the subject of boundary-setting in EIA. In the words of Noble (2000): "Defining the boundary within which to conduct an environmental impact assessment is often a challenging task. Perhaps the area of greatest concern, and ironically the area of least attention, is the definition of temporal boundaries in EIA".

In our view, collaborative boundary-setting through scoping continues to be an important aspect of focusing EIA studies. The setting of study boundaries may always be a challenge due to irreducible uncertainties surrounding the spatial and temporal dynamics of complex ecosystems, particularly in the context of a changing global climate. As with other elements of ecosystems identified during the scoping process, it may be best to consider initial study boundaries as being tentative. In this way, uncertain or fluctuating ecosystem boundaries may be adjusted for throughout the EIA process.

7.1.6 Drivers

Holling (1978) wrote that "In choosing variables we must be careful to distinguish between system state variables and driving variables". According to Holling (1978), state variables represent the ecological quantities we wish to predict, whereas as driving variables represent quantities known to influence the condition of state variables, but for which the model does not make predictions. Predictions about the future condition of system state variables are based largely on assumptions about the behaviour of system driving variables, which can represent environmental phenomena operating external to the ecosystem in question, or development actions imposed by humans onto that ecosystem. Holling (1978) suggested that one way to handle uncertainty in drivers (not to mention relationships and parameters) would be to construct what he called 'alternative models'. According to Holling (1978), such models would allow for the inclusion of potentially influential (yet highly uncertain) environmental factors.

Decades later, the issue of uncertainty surrounding external drivers, particularly global climate, has spurred on new conversations about the handling of uncertainty during the scoping stage of EIA processes. To accommodate uncertainties in ecosystem drivers (e.g., climate) and other external wildcards (e.g., induced development, market demand), several authors (e.g., Mulvihill 2003; Duinker and Greig 2007; Bond et al. 2015) have called for greater use of scenarios and scenario building in EIA. The use of such techniques, it is argued, necessarily opens up the scoping process to consider a wider range of speculative inputs, particularly those contextual factors that may be plausibly seen to influence future environmental conditions.

7.2 Ecological characterization

7.2.1 Overview

Foundational guidance materials for science in EIA (e.g., Holling 1978; Ward 1978; Munn 1979; Beanlands and Duinker 1983; Duinker and Baskerville 1986) emphasized the importance of establishing explicit, quantitative descriptions of ecological processes as a basis for subsequent experimentation, prediction, and monitoring efforts. According to this early literature, a process-based modelling exercise offers the following special benefits to EIA: (i) synthesis of ecological knowledge, (ii) transparency of assumptions, and (iii) a framework to guide empirical studies. Holling (1978) observed that once the underlying functional form of an ecological relationship had been established, additional information would be required for the estimation of: (i) parameters, (ii) initial state variables, and (iii) driving variables. For our purposes, the descriptive (i.e., characterization) stage of EIA is aimed at obtaining, interpreting, and synthesizing such information.

The earliest frameworks for conducting EIAs (e.g., Leopold et al. 1971; Dee et al. 1973) could depict neither ecological relationships nor spatiotemporal dimensions. Within such static and compartmentalized study frameworks, the 'count everything' approach to description inevitably prevailed. Indeed, early EIAs were typically characterized by the use of a one-time, comprehensive survey to identify and list all biophysical entities observed within a given study area. Subsequent commentaries and guidance materials (e.g., Holling 1978; Ward 1978; Munn 1979; Hilborn and Walters 1981) challenged EIA's preoccupation with comprehensive surveys and inventories. Ward (1978), for example, criticized two common approaches to environmental description that she labelled the 'busy taxonomist' approach and the 'information broker' approach. Similarly, Hilborn and Walters (1981) referred to traditional baseline and process studies in EIA as 'helicopter ecology'. They concluded that it was not exhaustive inventories or baselines that provided the key to understanding ecological systems, but the qualitative insight of a few experts into how ecosystem components and processes interact over space and time.

Holling (1978) pointed out that a more effective and efficient approach to characterizing ecosystems would be to establish functional relationships based on existing ecological knowledge. According to Holling (1978), "one can proceed farther than is normally thought possible in the face of meager data by mobilizing available insight into the system's constituent processes." Holling (1978) further stated that "as soon as we know that a particular mathematical function will describe a process, the information requirements are suddenly reduced greatly. Now we need only estimate values for the few parameters of that function." On the subject of description, Holling (1978) concluded that "the goal, then, of description is not description but useful explanation".

In outlining the fundamental purpose of effects monitoring in EIA, Duinker (1989) challenged the perceived need for so-called 'baseline studies', an often time-consuming and resource-intensive endeavour to produce exhaustive descriptions of environmental conditions prior to development. Duinker (1989) reasoned that the most common approach to defining an impact – a comparison of 'before' and 'after' measurements of environmental conditions – is based on the false assumption that such conditions are static (i.e., that they would have persisted in the absence of development). Likewise, he reasoned that a comparison of measurements taken in the ecosystem of interest with measurements taken in a similar ecosystem – another common approach to defining an impact – is based on the false assumption that both ecosystems are identical, and that one would have behaved exactly like the other in the absence of development.

Duinker (1989) concluded that the most logical approach to defining an impact would be to calculate the difference between two predicted time-series: one with the proposed development in place and one without. A time-series of data generated through monitoring would then serve as a check on one of the predicted time-series. Accordingly, Duinker (1989) argued that a statistically rigorous description of natural variability (i.e., 'baseline') – such as that advocated by Hirsch (1980) and Beanlands and Duinker (1983) – could only modestly assist in the process of impact prediction. He argued that because natural variation in the condition of a VEC is typically caused by several ecological processes, analysts should strive to understand how such processes contribute to the observed pattern of variability, rather than simply describe that pattern.

The literature of the 1990s and 2000s has not challenged the basic tenets of ecological characterization as described in the literature of the 1980s. Since that time, however, technological advancements in computational modelling (e.g., individual-based models) and remote/automated data acquisition (e.g., light imaging, detection, and ranging [LIDAR] and global positioning system [GPS] telemetry) have significantly expanded the range of tools, techniques, and datasets available for ecological characterization (e.g., Grimm 1999; Cooke et al. 2004; de Leeuw et al. 2010). The potential contributions of other kinds of ecological knowledge (e.g., local, traditional, Aboriginal) have also been the subject of considerable attention in the recent literature (e.g., Stevenson 1996; Usher 2000). In this section, we chart the evolution of the major tools and techniques available for ecological characterization in EIA.

7.2.2 Mobilizing science outside EIA

7.2.2.1 Functional relationships

At the time of EIA's inception in the early 1970s, ecological simulation modelling was just beginning to flourish. According to Jørgensen (2008), the earliest approaches to modelling represented three broad domains of ecology: (i) biogeochemical (i.e., flows of matter through ecosystems), (ii) bioenergetics (i.e., flows of energy through ecosystems), and (iii) population/community (i.e., growth and dynamics of wildlife populations/communities). The early literature on ecological modelling (e.g., May 1973; Maynard-Smith 1974; Pielou 1977) further subdivided such models based on the following characteristics: (i) state-dependent or time-dependent, (ii) continuous or discrete, and (iii) stochastic or deterministic. Additionally, population models could be assembled with or without age and class structure.

Early 'compartment' or 'food web'-type models were designed to simulate flows of carbon, energy, and nutrients through different kinds of ecosystems, including grasslands (e.g., Patten 1971; Innis 1975), forests (e.g., Shugart et al. 1974; O'Neill 1975; Overton 1975), deserts (e.g., Goodall 1975), tundra (e.g., Miller et al. 1975), and lakes (e.g., Walters and Efford 1972; Park 1974).

Early population simulation models were designed to simulate a range of predator-prey or animal-habitat interactions. Examples included a model for simulating the dynamics of a caribou-lichen system (Walters et al. 1975), a model for simulating the dynamics of a wolf-moose system (Zarnoch and Turner 1974), and a model for simulating the dynamics of a lamprey-trout system (Lett et al. 1975).

Early community simulation models included the so-called forest gap models (e.g., Botkin et al. 1972; Ek and Monserud 1974; Shugart and West 1977), designed to simulate the successional dynamics of small forest stands following an opening in the canopy.

The 1970s also saw the widespread use of hydrological models, which were designed to simulate the physical and chemical dynamics of ground water (e.g., Freeze 1971; Winter 1978) and surface water (e.g., Quick and Pipes 1976; Solomon and Gupta 1977). Likewise, a range of geomorphological models were designed to simulate the dynamics of erosion and sedimentation processes (e.g., Onstad and Foster 1975; Bridge and Leeder 1979).

Finally, the 1970s saw the application of general systems theory to ecological modelling, which aimed to integrate the various sub-disciplines outlined above. Walters (1974), for example, described an interdisciplinary approach to watershed modelling, whereby hydrodynamics, vegetation growth, and a variety of fish and wildlife populations were functionally linked together.

At the time of the Beanlands and Duinker (1983) report, scientists were beginning to focus more explicitly on the spatial dimension of complex ecosystems, as evidenced by the emerging field of landscape ecology (e.g., Forman and Godron 1981; Wiens et al. 1985). According to the literature (e.g., Huston et al. 1988; Baker 1989), advances in computational power, coupled with the advent of GIS and satellite remote sensing, allowed for the development of individual-based and spatially explicit simulation models. Huston et al. (1988) argued that such models would foster greater collaboration and synthesis among the various scientific disciplines. Early examples included a model for simulating hydrological flows and vegetation succession in a forested floodplain (Pearlstine et al. 1985), a model for simulating the physical transportation of water and sediments in a coastal marsh (Costanza et al. 1988; Sklar et al. 1985), and a model for simulating fire and post-fire succession in wet or semi-arid forests (Kessell et al. 1984).

With growing interest in Earth's global biogeochemical cycles, the 1980s also saw the development of numerous models for linking cross-scale atmospheric, marine, and terrestrial processes. Examples included models for simulating the biogeochemical and successional dynamics of coupled soil-vegetation systems (e.g., Solomon 1986; Pastor and Post 1988), models for simulating the biogeochemical and climatic dynamics of coupled vegetation-atmosphere systems (e.g., Dickinson and Henderson-Sellers 1988; Sellers et al. 1986; Running and Coughlan 1988), and models for simulating the biogeochemical, hydrological, and successional dynamics of forests, lakes, and watersheds under acid precipitation regimes (e.g., Booty and Kramer 1984; Cosby et al. 1985; Nikolaidis et al. 1988).

The literature of the 1990s and 2000s has since revealed widespread interest in the development of individual-based and spatially explicit simulation models. Several spatially explicit landscape models have been developed for simulating the productivity and successional dynamics of broad-scale forest landscapes (e.g., Scheller and Mladenoff 2004; Scheller et al. 2007). Likewise, a number of spatially explicit population models have been developed for simulating the dispersal and survival of wildlife populations in patchy landscapes (e.g., Mckelvey et al. 1992; Pulliam et al. 1992; Dunning et al. 1995). Similar models have also been developed for freshwater fish populations (e.g., Van Winkle et al. 1998).

To incorporate more realistic habitat dynamics, a number of authors have proposed frameworks for linking spatially explicit population models with forest landscape models using habitat suitability indices (e.g., Akçakaya 2001; Larson et al. 2004; McRae et al. 2008; Franklin 2010). Such approaches to modelling, however, have been controversial, as the rules governing animal dispersal, movement, and mortality remain highly sensitive to parameter adjustments (see Ruckelshaus et al. 1997; Mooij and DeAngelis 1999). At the same time, other authors (e.g., Morrison et al. 2006) have cautioned that such models do not consider important processes such as predation and interspecific competition, which can have a strong influence on population dynamics. Still other authors (e.g., Railsback and Harvey et al. 2002) have observed that the simplistic rules used to simulate animal movement often fail to reproduce realistic patterns of behaviour.

More defensible, systems analytical approaches to simulation modelling seem to have had lasting appeal in the scientific literature. For terrestrial wildlife, several population models have been developed to simulate the dynamics of predator-prey-habitat systems (e.g., Weclaw and Hudson 2004). For marine life, fisheries scientists have found it more useful to rely on population models linked to trophic compartment-type models (e.g., McClanahan 1995; Walters et al. 1999; Christensen and Walters 2004). This, together with ongoing interest in modelling the carbon, water, and nutrient budgets of terrestrial ecosystems (e.g., Billen and Garnier 1999; Kurz et al. 2009), suggest that perhaps not much has changed in the fundamentals of ecological modelling since the 1980s.

7.2.2.2 *Parameters and process studies*

According to the scientific literature (e.g., Richter and Sondergrath 1990; Hilborn and Walters 1992; McCallum 2000), parameters represent a crucial link between data and models, or between empirical and theoretical science. It is also widely recognized (e.g., Jørgensen and Bendoricchio 2001; Soetaert and Herman 2009) that the most consistent and reliable parameter estimates are derived from the published results of long-term empirical studies. Different kinds of models will require different kinds of parameters, but the following general types can be delineated: (i) life-history (e.g., Hindell 1991; Wich et al. 2004), (ii) demographic (e.g., Fancy et al. 1994; Best et al. 2001), (iii) physiological (e.g., Sullivan et al. 1996), (iv) biogeochemical (e.g., Soer 1980; Kavvadias et al. 2001), (v) hydrological (e.g., Reimers 1990; Zavattaro and Grignani 2001), and (vi) disturbance (e.g., Anderson et al. 1987; Seymour et al. 2002). Many helpful guides to parameterizing and calibrating simulation models can be found in the literature (e.g., Janssen and Heuberger 1995; Jørgensen and Bendoricchio 2001). Here we chart the general development of empirical biophysical science since the 1970s, as described in the formal peer-reviewed literature. In doing so, we attempt to highlight major technological and methodological advancements related to the study of ecological processes and parameters.

At the time of EIA's inception in the early 1970s, ecological research was dominated by short time frames and a focus on local-scale sites (CEQ 1974; Callahan 1984). For wildlife studies, many scientists (e.g., Dolbeer and Clark 1975; Wilbur 1975) relied on the use of traditional mark-recapture and trapping techniques to estimate demographic parameters, while other scientists (e.g., Cook et al. 1971; Carroll and Brown 1977) had begun to use very high frequency (VHF) radio telemetry. In addition to population studies, radio telemetry could be used to characterize animal home ranges (e.g., Van Ballenberghe and Peek 1971; Trent and Rongstad 1974), habitat use (e.g., Craighead and Craighead 1972; Nicholls and Warner 1972), and physiological state (e.g., Priede and Young 1977; MacArthur et al. 1979).

To characterize biogeochemical flows of matter and energy between ecosystem 'compartments', many scientists (e.g., Schlesinger 1977; Whittaker et al. 1979) relied on traditional field and laboratory techniques such as isotopic labelling, chromatography, spectrophotometry, elemental analyzers, and gas exchange methods. Meanwhile, other scientists (e.g., Kelly et al. 1974) had begun to profit from the use of automated, in-situ measurement devices such as electrochemical sensors.

In terms of broad-scale, collaborative research, the 1970s saw the end of the International Biological Program (IBP), a decade-long, multi-site research effort focused on the study of ecological productivity (NAS 1975). Despite its many shortcomings, the IBP would make several important contributions to the study of biogeochemical processes, particularly with respect to soil nutrients, organic decomposition, and plant productivity (e.g., Gosz et al. 1973; Whittaker et al. 1979).

Finally, the 1970s saw the launch of the first land remote sensing satellites (Landsat 1–3), whose continuous stream of data would be used, amongst other things, to characterize the photosynthetic properties of tree canopies (e.g., Rouse et al. 1973; Tucker 1979). The initiation of the Landsat program in the 1970s marked the beginning of a new era in global, space-based monitoring systems.

In the 1980s, the field of ecology expanded to consider functional processes operating at diverse spatial and temporal scales (e.g., Holling 1986). To foster greater continuity and collaboration in ecological research, the scientific and policy communities came together in 1980 to establish the Long-Term Ecological Research (LTER) network, a collaborative, multi-site research program based on standardized methods of observation and data storage (e.g., Likens 1983; Callahan 1984, Webster et al. 1985). Throughout the 1980s, the LTER program would begin characterizing a range of ecological processes and parameters within different biomes, with a particular emphasis on nutrient cycling, primary productivity, and forest succession (e.g., Knapp and Seastedt 1986; Mattson et al. 1987b; Magnuson et al. 1990).

To advance the study of Earth's global biogeochemical systems, and to gain a better understanding of the role of humans in altering those systems, the international scientific community came together in 1980 to establish the World Climate Research Program (WCRP), and again in 1986 to establish the International Geosphere Biosphere Program (IGBP) (e.g., Schiffer and Rossow 1983; NRC 1986b). Throughout the 1980s, the WCRP and IGBP would together rely on space-based remote sensing, in coordination with ground-based monitoring efforts (e.g., LTER), to characterize a range of cross-scale biogeochemical processes linking Earth's atmosphere, oceans, and land surfaces (e.g., Sellers et al. 1988; Goward and Hope 1989; Houghton et al. 1990).

In a related set of developments (e.g., NASA 1984), significant improvements were made in high resolution and multi-spectral remote sensing (e.g., NOAA-AVHRR, Landsat-TM). These new instruments, along with improved analytical techniques, would allow for the estimation of more-accurate model parameters, particularly those associated with tree canopy photosynthesis and productivity (e.g., Holben 1986; Rock et al. 1986; Huete 1988).

Finally, the late 1980s saw the application of Argos – a satellite-based weather monitoring system – to the tracking of free-ranging wildlife (Fancy et al. 1988). Unlike traditional radio transmitters, which were limited by their short signal range, Argos-based systems were capable of tracking animal migrations over vast distances (e.g., Strikwerda et al. 1986; Fancy et al. 1989). By coupling these data with simultaneous remote sensing measurements in a GIS, broad-scale animal movements could be linked to dynamic environmental conditions and habitat characteristics (e.g., Priede 1984). The literature of the 1990s and 2000s has since highlighted many new technologies with potential applications in empirical biophysical science. New satellite-borne remote sensing instruments (e.g., moderate resolution imaging spectroradiometer [MODIS]) offering greater spectral coverage and resolution have provided new and improved datasets for use in ecological studies (Justice et al. 1998). Noteworthy applications include the characterization of wildfire regimes (Justice et al. 2002), enhanced vegetation indices (Huete et al. 2002), and global land cover mapping (Friedl et al. 2002). Recently, aircraft-borne LIDAR sensors have emerged as a viable option for characterizing the distribution, three-dimensional structure, and functional characteristics of forest stands (Lefsky et al. 2002).

Likewise, several technological advancements in satellite telemetry (e.g., GPS-based systems, miniaturized tags, coded signals, lithium batteries, archival loggers) have allowed for more-reliable and more-accurate animal tracking studies (Rodgers 2001; Cooke et al. 2013). Tracking devices that provide information on an animal's physiology, behaviour, and energetic status have also recently emerged as a viable option for characterizing interactions between individual animals and their environment (Cooke et al. 2004). Finally, the 1990s and 2000s have seen the establishment of automated, wireless sensor networks for monitoring a variety of biogeochemical parameters, fluxes, and state variables at a variety of spatial and temporal scales (Collins et al. 2006; Jones et al. 2010). When simultaneously integrated with adjacent sensor networks, remote sensing platforms, and GPS/Argos telemetry systems, these multi-tiered monitoring platforms have the potential to generate powerful insights into the relationships linking local animal behaviour, regional environmental conditions, and global-scale biogeochemical drivers (Rundel et al. 2009; Handcock et al. 2009).

In summary, empirical biophysical science has changed significantly since the time of EIA's inception in the early 1970s. Driven primarily by advances in theory and technology, local-scale biophysical science has been progressively expanded to regional and global scales through integration with Earth systems science. The long-term monitoring efforts associated with these developments have resulted in more robust characterizations of ecological parameters, as well as a greater understanding of the dynamic processes and feedbacks linking global climatic and biogeochemical changes to local-scale ecosystems, communities, and populations.

7.2.3 Data collection inside EIA

7.2.3.1 Field surveys

While long-term ecological research is thought to provide the most consistent and reliable parameter estimates, site-specific studies are still needed inside EIA to characterize the initial conditions of VECs and other related variables. Considering the limited time frames in which most EIAs must operate, Holling (1978) emphasized the importance of accessing existing inventories and databases wherever possible, including those maintained by government agencies and industry groups. Ward (1978), on the other hand, outlined some basic approaches for designing and implementing short-term field surveys in EIA. She described the use of mark-recapture techniques to estimate wildlife abundances and the use of field sampling and analytical chemistry techniques (e.g., elemental analyzers, spectrophotometry, chromatography) to characterize physical media such as water and soils. Subsequent guidance materials published in the 1980s (e.g., Beanlands and Duinker 1983; Westman 1985; NRC 1986a) described the use of newer tools and techniques (e.g., satellite remote sensing, radio telemetry) to characterize important wildlife habitat.

Decades later, the subject of field surveys has not received much attention in the EIA literature. A quick search of the scientific literature, however, reveals several detailed protocols for the characterization of air (e.g., Chow 1995; Baron and Willeke 2001), water (e.g., Hounslow 1995; Fetter 2000), soils (e.g., Carter and Gregorich 2008), terrestrial carbon stocks (e.g., Brown 2002; Gibbs et al. 2007), wildlife populations (e.g., Seber 1982; Buckland et al. 2015), and wildlife habitats (e.g., Manly et al. 2002).

7.2.3.2 Local, traditional, and aboriginal knowledge

Recently, the potential application of other kinds of ecological knowledge (local, traditional, Aboriginal) has been the subject of considerable discussion in the literature (e.g., Freeman 1992; Johannes 1993; Sallenave 1994; Stevenson 1996; Huntington 2000; Usher 2000). These authors delineate some major categories of informal ecological knowledge, the overall applicability of such knowledge to EIA processes, and some of the major challenges associated with its collection, interpretation, and integration with formal scientific knowledge. According to Stevenson (1996), many Aboriginal resource users have had extensive contact with their local environment, and therefore possess a richer and more detailed understanding of it than most outsiders do. Stevenson (1996) explained that based on their familiarity with local environments, Aboriginal resource users can, if they so choose, contribute valuable insights to EIA studies, particularly with respect to wildlife and wildlife habitat. In a similar vein, Usher (2000) noted that the substantial time-depth of traditional observations promises to provide better explanations for existing environmental conditions than descriptions derived from short-term field surveys, particularly in remote and poorly studied ecosystems.

Both Stevenson (1996) and Usher (2000) pointed out that traditional ecological knowledge studies often require far less time and money to conduct than conventional baseline studies. They also concluded that for traditional and Aboriginal knowledge to realize its full potential in EIA, knowledgeable elders and resource users must be encouraged to play an active role in determining how their knowledge will be documented, interpreted, and used within the EIA process. In absence of such collaborative arrangements, there is a risk that such knowledge may be misinterpreted, misused, and even exploited by non-Aboriginal interests.

7.2.3.3 Integrating field surveys and traditional knowledge

The recent literature provides numerous examples of how scientific and traditional observations can be integrated to explain complex ecological phenomena, particularly in northern and remote environments (e.g., Huntington et al. 2004a, 2004b; Gilchrist et al. 2006; Gagnon and Berteaux 2009). According to Gagnon and Berteaux (2009), traditional ecological knowledge "has generally been recognized as differing from science because it is based on information acquired during longer time series but over smaller and more specific localities." Considering their differences, Huntington et al. (2004a) observed that both kinds of knowledge may be mutually exchanged and integrated to understand complex, multi-scale phenomena.

Summarizing recent collaborations with resource users in the Arctic, Huntington et al. (2004b) explained how satellite telemetry data describing the broad-scale migration patterns of duck and whale populations were paired with the traditional ecological knowledge of local hunting communities to explain recently observed population declines. In one example, telemetry data showed that ducks breeding in the Northwest Territories during the summer months migrated each year to nesting and moulting areas in Alaska during the winter months. Huntington et al. (2004b) concluded that the timing and migratory connection between the two distant areas suggests that locally observed population declines in the Northwest Territories could be attributed to a viral disease outbreak in Alaska that caused the collapse of a local herring population (an important food source for migratory waterfowl).

In a similar example, Gilchrist et al. (2006) showed how local and formal scientific knowledge could be integrated to explain locally observed duck population declines in Nunavut. They concluded that the timing and circumstances surrounding a particular mass die-off event suggest links between a volcanic eruption in the Philippines, subsequent atmospheric cooling in the Polar Regions, and persistent winter ice that could prevent feeding by the ducks in open water polynyas. Such case studies provide strong evidence for the integration of traditional and formal scientific knowledge inside EIA, particularly when developments are proposed in remote and poorly studied ecosystems.

7.3 Cause-effect research

7.3.1 Need for cause-effect knowledge inside EIA

Holling (1978) maintained that long-term observational studies, though they provide the data needed to establish functional process relationships and initial parameter estimates, only reflect the behaviour of ecological systems within a narrow range of historical or naturally occurring conditions. Holling (1978) reasoned that because ecosystems typically display complexities in their response to disturbance events (e.g., nonlinearities, time lags, thresholds), extrapolation beyond observed conditions cannot reliably predict the effects of novel development actions. Consequently, he and others (e.g., Ward 1978; Hilborn and Walters 1981; Suter 1982; Beanlands and Duinker 1983; NRC 1986a; Walters and Holling 1990; Walters 1993) advocated the use of carefully designed laboratory, field, and ecosystem experiments to reduce uncertainty surrounding the effects of development actions.

According to Holling (1978), understanding how an ecological parameter or variable of interest (e.g., reproductive rate, survival rate) might respond to a particular development action would require the following steps: (i) disaggregation of that parameter or variable into its constituent components/processes (e.g., mating, nesting, food supply), and (ii) design of an experiment or set of experiments to study the effects of proposed development actions on each of those components. Holling (1978) also noted, however, that due to the broad spatial and temporal scales over which many development-related impacts occur (e.g., hydroelectric dams, bioaccumulation of toxins), it may be difficult or even impossible to design and conduct the necessary experiments. Accordingly, he and others (e.g., Ward 1978; Beanlands and Duinker 1983; NRC 1986a; Walters 1986) emphasized the importance of treating the development itself as an experiment. In this way, anticipated effects would be predicted prior to development, actual effects monitored during implementation, and new knowledge created to support predictive endeavours in the future.

Recently, Greig and Duinker (2011) have observed that the short time frames in which most EIAs are conducted generally preclude the use of large-scale perturbation experiments, which themselves require ample time and resource to conduct. Such experimentation, they argue, is better suited to science outside EIA, which is well equipped to provide scientists inside EIA with the resulting effects knowledge. To complete the circle, science inside EIA must then predict and monitor the effects of development, thereby testing and refining the general knowledge provided by science outside EIA. Here we chart the evolution of experimental cause-effect research conducted outside EIA, as described in the scientific literature published since the early 1970s.

7.3.2 Creation of cause-effect knowledge outside EIA

Generally speaking, most cause-effect experiments conducted in the 1970s were aimed at understanding the demographic, behavioural, physiological, hydrological, and biogeochemical consequences of pollution. Examples included the use of laboratory bioassays to study the effects of water pollution on fish (e.g., Leach and Thakore 1975; Hara et al. 1976), field plots and fumigation chambers to study the effects of air and soil pollution on plants (e.g., Shure 1971; Hill et al. 1974; Koterba et al. 1979), and loud equipment or noise-making devices to study the effects of noise pollution on wildlife (e.g., Freddy et al. 1977; White et al. 1979). Several small experiments were also conducted to study the effects of thermal pollution and mechanical entrainment/impingement on young fish, eggs, and larvae drifting or swimming near thermal power plants (e.g., Rulifson 1977; Dorn et al. 1979).

In addition to simple laboratory and field experiments, the 1970s saw widespread interest in modelling the effects of pollution using artificially simplified ecosystems. Examples included indoor microcosms for studying the effects of soil pollution on forests (e.g., Jackson et al. 1978; Ausmus et al. 1978), and outdoor microcosms for studying the effects of water pollution on marine biota (e.g., Evans 1977). Somewhat larger and more ambitious research initiatives relied on outdoor 'mesocosms' to simulate the effects of pollution on marine ecosystems (e.g., Menzel and Case 1977; Kremling et al. 1978).

Finally, the 1970s saw some of the first whole-ecosystem perturbation experiments, which were mostly aimed at understanding the causes and effects of water pollution. Examples included large-scale experiments for studying the effects of fertilization and nutrient enrichment on entire lakes (e.g., Schindler et al. 1971), and watershed experiments for studying the effects of deforestation and herbicide application on stream water (e.g., Likens et al. 1970).

In the 1980s, large ecosystem experiments expanded to accommodate growing concerns over the effects of acid rain. Examples included the artificial acidification of streams, lakes, and wetlands (e.g., Hall et al. 1980; Schindler et al. 1980; Bayley et al. 1987), and the use of glasshouses to study the effects of acid deposition on forests (e.g., Wright et al. 1988).

In addition to ongoing pollution research, the 1980s saw some of the first large experiments aimed at characterizing the effects of species removals, species additions, and habitat disturbances. Examples included whole-lake experiments for studying the effects of fish population removals and additions on lake productivity and nutrient cycling (e.g., Carpenter and Kitchell 1988; Carpenter et al. 1988); experiments for studying the effects of aircraft noise on wildlife habitat-use (e.g., Krausman et al. 1986); wetland experiments for studying the effects of water level manipulation on vegetation succession and waterfowl habitat selection (e.g., Murkin and Kaldec 1986; van der Valk et al. 1994; Murkin et al. 1997); and grassland experiments for studying the effects of mowing, burning, and cattle grazing on the abundance, diversity, and productivity of prairie vegetation (e.g., Abrams and Hulbert 1987; Collins 1987; Gibson 1989).

The 1980s also saw the initiation of the first large habitat fragmentation experiments. By deliberately cutting and burning swaths of tropical rainforest, such experiments would begin to understand the effects of habitat fragmentation on species diversity, population demographics, and local microclimate in remnant forest patches of various sizes (e.g., Malcolm 1988; Rylands and Keuroghlian 1988; Kapos 1989; Bierregaard and Lovejoy 1989).

More recently, cause-effect experiments have again expanded to accommodate growing concerns over the effects of climate change. Examples include the use of glasshouses to study the effects increased temperature and CO_2 on forests (e.g., van Breemen et al. 1998; Wright 1998); the use of free air CO_2 enrichment technology to study the effects of increased temperature and CO_2 on forests, grasslands, and deserts (e.g., Hendrey et al. 1999; Jordan et al. 1999; Reich et al. 2001); and the use of field enclosures and mesocosms to study the effects of nutrient enrichment, acidification, and elevated CO_2 on coral reefs (e.g., Koop et al. 2001; Langdon et al. 2003).

In addition to climate change concerns, several emerging industries and technologies have posed new and potentially unacceptable environmental consequences (e.g., transgenic crops, aquaculture, seabed mining, hydraulic fracturing). Though the environmental impacts of these emerging industries were largely unknown a few decades ago, experimental research has been successful in reducing some of the critical uncertainties. Examples include the use of laboratory, field, and watershed experiments to study the effects of hydraulic fracturing and shale gas development on water, soil, and forest vegetation (e.g., Murdoch 1992; McKay et al. 1993; Adams 2011; McBroom et al. 2012); the use of laboratory, field, and glasshouse experiments to study the effects of genetically modified crops on soil ecosystems (e.g., Birch et al. 2007; Cortet et al. 2007); and the use of wholelake experiments and pilot studies to examine the effects of fish farming on freshwater and marine ecosystems (e.g., Findlay et al. 2009; Rooney and Podemski 2009). Likewise, concern over the increasing exploitation of seabed resources has led to a range of benthic impact experiments. Examples include the use of experimental trawling and dredging to study the effects of fishing gear and aggregate extraction on marine benthos (e.g., Kenny and Rees 1996; Bradshaw et al. 2001; Kenchington et al. 2006) and experimental seabed disturbances to study the effects of deep sea mining on marine ecosystems (e.g., Thiel et al. 2001; Sharma 2001).

As pointed out by Greig and Duinker (2011), science outside EIA has demonstrated its ability to design and implement rigorous, long-term experiments in large ecosystems. The scientific enterprise inside EIA, which is often faced with severe time restrictions, could therefore benefit substantially by drawing upon such external resources. At the same time, science outside EIA could also benefit from having relevant effects knowledge tested and refined through site-specific application inside EIA. In sum, both scientific communities, through ongoing collaboration and experimentation, could mutually support one another in the common pursuit of reliable knowledge and sustainable development decisions.

7.4 Impact prediction

7.4.1 Prediction and uncertainty in EIA

For our purposes, the terms 'forecast', 'projection', and 'prediction' are treated as synonyms. Each connotes a statement specifying the expected future condition of a thing of interest. In EIA, the things of interest are specified components of the environment, including both objects (e.g., wildlife, soil nutrient pools) and processes (e.g., predation, nutrient cycling). Predictions are always conditional statements and are stated in terms of the units of measurement used to define the starting condition of the thing of interest. They take the following general form: if specific relationships and starting conditions hold, then the thing of interest is expected to have the following conditions through a specified future time at specified locations. In EIA, the relationships and starting conditions must include the undertaking being assessed.

The earliest frameworks for EIA (e.g., Leopold et al. 1971; Dee et al. 1973) treated the notion of impact prediction trivially, relying on pseudo-quantitative ranking and scoring methods to convey the opinion of analysts and experts. Subsequent commentaries and guidance materials for EIA (e.g., Andrews 1973; Carpenter 1976; Schindler 1976; Holling 1978; Ward 1978; Munn 1979) criticized earlier methods, calling for more rigorous approaches to impact prediction based on ecological knowledge, experimentation, and simulation modelling. Recognizing the fundamental uncertainty surrounding complex ecosystems, Holling (1978) maintained that it would be impossible to predict the environmental impacts of development with absolute certainty. He argued, therefore, that the purpose of impact prediction should be to explicitly identify and reduce key uncertainties during the early stages of development planning and design. In this way, critical uncertainties might have a better chance of being effectively managed and reduced through ongoing monitoring and follow-up.

In addition to experimentation, Holling (1978) outlined several techniques that might be used to test predictions and characterize lingering uncertainties prior to decision-making. First, he outlined a process of 'invalidation', in which a model's predictive outputs are compared with the effects of past perturbations in similar ecosystems, thereby testing and even expanding the limits of a model's credibility. Second, he described a process of sensitivity analysis, in which a model's fixed parameters are adjusted individually and then simultaneously to test the responsiveness of predictions to parameter uncertainty. Lastly, he described a process of generating alternative models, in which analysts explore the consequences of adding or removing potentially influential (yet extremely uncertain) processes or driving variables. According to Holling (1978), the end result would be a small set of plausible models that consider all potentially relevant factors, provide reliable impact predictions, yet explicitly preserve remaining uncertainties for ongoing management and reduction.

During the 1980s, several commentaries and reviews (e.g., Rosenberg et al. 1981; Caldwell et al. 1982; Beanlands and Duinker 1983; Clark et al. 1983; Culhane 1987) continued to highlight major shortcomings in formal EIA practice, particularly with respect to impact predictions, which were found to be vague, untestable, or otherwise non-existent. In a Canadian context, Beanlands and Duinker (1983) outlined the following general standards for impact prediction in EIA: (i) predictions should be stated explicitly in testable, quantitative terms; (ii) the timing and spatial extent of predicted impacts should be specified; (iii) the range of uncertainty in impact predictions should be specified; and (iv) the conceptual and technical basis for impact predictions should be made clear.

Duinker and Baskerville (1986) later expanded on these general standards by outlining a somewhat more detailed framework for impact prediction in EIA. Among other things, Duinker and Baskerville (1986) highlighted the following as core elements of a rigorous, scientific approach: (i) quantitative, process-based, feedback-type models; (ii) explicit connections between development actions and environmental components of interest; (iii) a minimum of two predictions (one with the proposed development in place and one without); (iv) an appropriate level of spatial and temporal resolution; and (v) use of sensitivity analyses to characterize remaining uncertainties. Together with earlier guidance materials, the principles and protocols outlined by Beanlands and Duinker (1983) and Duinker and Baskerville (1986) established a strong basis for making useful and defensible impact predictions in EIA.

While the topic of environmental impact prediction has continued to receive attention in the literature of the 1990s and 2000s (e.g., Armstrong 1999; Clark et al. 2001; Demyanov et al. 2006), the basic standards for prediction outlined in the literature of the 1980s have not been challenged or expanded. Cashmore (2004), however, has recently questioned the role of science in EIA, arguing for a re-conception of EIA as a largely political process. In response, Greig and Duinker (2011) have argued for an ongoing pivotal role for science in EIA – that is, to provide useful and defensible predictions of environmental impact.

In a similar vein, the issue of uncertainty (and how best to deal with it) has been a topic of considerable discussion in the recent EIA literature, particularly with respect to modelling and impact prediction. Reviews of theory (e.g., Refsgaard et al. 2007; Bond et al. 2015; Leung et al. 2015; Cardenas and Halman 2016) and practice (e.g., Tennøy et al. 2006; Duncan 2008; Wardekker et al. 2008; Larsen et al. 2013; Leung et al. 2016; Lees et al. 2016) have together highlighted major shortcomings in the characterization, disclosure, and ongoing reduction of scientific uncertainty in EIA.

To improve explorations of the future, Duinker and Greig (2007) recently proposed greater use of scenarios and scenario analysis in EIA. They argue that scenario analysis provides a crucial means of grappling with uncertainties inherent in the prediction of environmental impacts. According to Duinker and Greig (2007), the minimum of two predictions needed to define an environmental impact (i.e., one with the proposed development in place and one without) should be expanded to include a range of plausible future scenarios that explore the consequences of potentially influential (yet highly uncertain) driving variables.

The prediction of environmental impacts, along with the explicit communication and reduction of scientific uncertainty, continue to be recognized as core elements of EIA's contribution to sustainable development. In the following section we review the application of science to environmental impact prediction, as described in the peer-reviewed scientific literature published since the 1970s. Although our review of the literature focuses on the science of environmental impact prediction outside formal EIA, some examples of regulatory practice have found their way into the literature reviewed here.

7.4.2   Using science to predict environmental impacts

In the 1970s, a wide range of hydrological models were used to predict the effects of point-source pollution on water quality. Examples included a model for predicting the effects of highway de-icing salts on groundwater quality (Gelhar and Wilson 1974), a model for predicting the effects of mine drainage on stream water quality (Herricks et al. 1975), and a model for predicting the effects of sewage discharges on lake water quality (Canale et al. 1973). Likewise, a range of atmospheric dispersion models were used to predict the effects of point-source pollution on air quality. Examples included a model for predicting the effects of airport traffic on local air quality (Daniels and Bach 1976), and a model for predicting the downwind effects of thermal power plant emissions (Carpenter et al. 1971). Dispersion models were also used to predict the physical transport and fate of marine oil spills (Blaikley et al. 1977).

In addition to predicting changes in air and water, several models were used to predict the effects of industrial food and timber production on soils. Examples included a model for predicting the effects of agricultural irrigation on soil nitrogen and phosphorus pools (Dutt et al. 1972), and a model for predicting the effects of timber production on forest floor organic matter and nitrogen pools (Aber et al. 1978). The 1970s also saw the use of simulation models to predict the effects of pollution, harvesting, and physical infrastructure on fish and wildlife populations. Examples included a model for predicting the effects of power plant cooling systems on freshwater fish populations (DeAngelis et al. 1977); a model for predicting the effects of exploitation and enhancement on wild salmon populations (Peterman 1975); and a model for predicting the effects of pollution, harvesting, and hydroelectric development on fish and wildlife populations (Walters 1974).

By the 1980s, EIA scholars and practitioners (e.g., Beanlands et al. 1985; Peterson et al. 1987; Sonntag et al. 1987; Stakhiv 1988) had begun to emphasize the importance of predicting cumulative environmental effects – that is, the effects of multiple natural and anthropogenic stressors acting simultaneously. Indeed, several simulation models were used to explore the environmental consequences of multiple development activities. Examples included a model for predicting the effects of harvesting, stocking, and lamprey reduction on trout populations (Walters et al. 1980); a model for predicting the effects of fire and wildlife management on tree canopy regeneration (Pellew 1983); a model for predicting the effects of exploitation and water pollution on freshwater fish populations (Goodyear 1985); and a model for predicting the effects of low-flying aircraft and coyote predation on a threatened pelican population (Bunnell et al. 1981).

In addition to more complex models, simpler models were used to predict the individual environmental effects of development actions. Examples included models for predicting the effects of oil spills on marine fish and mammal populations (Spaulding et al. 1983; Reed et al. 1984; Reed et al. 1989); models for predicting the effects of power plant cooling systems on marine and freshwater fish populations (Jensen 1982; Shuter et al. 1985; Polgar et al. 1988); and models for predicting the effects of exploitation on the growth and productivity of forests, fish, and wildlife resources (Deriso 1980; Shugart et al. 1980; Williams 1981).

The 1980s also saw the first use of spatially explicit simulation models for predicting the effects of human developments on terrestrial landscapes. Examples included a model for predicting the effects of hydroelectric development on water flows and vegetation succession in a forested floodplain (Pearlstine et al. 1985) and a model for predicting the effects of canals, levees, and sea level rise on water and sediment transport in a coastal wetland (Sklar et al. 1985).

In the 1990s and 2000s, the science of environmental impact prediction expanded greatly, with a growing emphasis on the broad-scale consequences of land-use change, habitat fragmentation, and global climate change. For instance, dynamic landscape models have been used to predict the effects of industrial timber harvesting and climate change on the long-term productivity and composition of forested watersheds (e.g., Scheller and Mladenoff 2005; Steenberg et al. 2011). Such models have also been linked with wildlife population models to predict the effects of timber harvesting and climate change on the dispersal and survival of threatened animal populations (e.g., Larson et al. 2004; McRae et al. 2008). In terms of biodiversity, forest landscape models have been used to predict the effects of timber production and other management practices on remaining habitat structure and suitability (e.g., Marzluff et al. 2002; Shifley et al. 2006; Gustafson et al. 2007). In a similar vein, a variety of carbon budget models have been used to predict the effects of timber production, climate change, and insect outbreaks on terrestrial carbon stocks (e.g., Karjalainen et al. 2002; Seidl et al. 2008; Kurz et al. 2008).

At the strategic level, various hydrological and soil process models have been used to predict the effects of land-use and climate change on the water and nutrient budgets of watersheds (e.g., Ferrier et al. 1995; Niehoff et al. 2002; Mango et al. 2011). Likewise, a variety of hydrological and microclimatic models have been used to predict the effects urban development on local water and energy budgets (e.g., Jia et al. 2002; Rosenzweig et al. 2009). For marine environments, coupled ocean-biogeochemical-population models have been used to predict the effects of exploitation and climate change on the productivity of fish stocks (e.g., Lehodey et al. 2013; Lindegren et al. 2010). Similar models have also been used to predict the effects of fish harvesting and climate change on the productivity of coupled fish-coral ecosystems (e.g., McClanahan 1995; Sebastián and McClanahan 2012).

In terms of new and emerging industries, a variety of simulation models have been used to predict the effects of fish farms on marine and freshwater quality (e.g., Håkanson and Carlsson 1998; Wild-Allen et al. 2010), the effects of shale gas development on groundwater quality (e.g., Schwartz 2015; Reagan et al. 2015), and the effects of wind energy developments on migratory bird populations (e.g., Eichhorn et al. 2012; Bastos et al. 2015). Still, an immense variety of simulation models have been used to predict the environmental impacts of more traditional developments. Examples include models for predicting the effects of roads and other noisy infrastructure on wildlife populations (e.g., Weclaw and Hudson 2004), the effects of mining and mineral processing on air and water quality (e.g., Herr et al. 2003; Dutta et al. 2004), and the effects of dams and artificial reservoir management on fish populations (e.g., Jager et al. 1997; McCleave 2001).

As pointed out by Greig and Duinker (2011), science outside EIA has developed the capacity to generate powerful insights into ecological processes and cause-effect relationships. Indeed, the scientific literature describes the development and application of an abundance of process-based models for environmental impact prediction. In our view, the scientific enterprise situated inside EIA could benefit substantially by calling in such knowledge. Likewise, the scientific enterprise situated outside EIA could benefit from having such knowledge applied and refined through site-specific application inside EIA.

7.5 Impact significance determination

7.5.1 Overview

In EIA, significance refers to the overall acceptability, legality, or sustainability of predicted environmental impacts, typically in terms of societal norms, regulatory standards, ecological thresholds, or some combination of the three (e.g., Lawrence 2007a, 2007b, 2007c). Ultimately, the significance attributed to predicted impacts will play a key role in determining whether a proposed development will receive final regulatory approval. The determination of impact significance is therefore closely linked to regulatory and environmental decision-making. While the bulk of scientific work in EIA is directed at generating reliable impact predictions, the determination of impact significance represents the critical stage at which the acceptability of future VEC conditions is judged. As such, significance evaluation is arguably one of the more challenging and disputed aspects of an EIA. According to Lawrence (2007a), there are two general approaches to impact significance determination: technical and collaborative. A third approach, called reasoned argumentation, seeks to integrate technical and collaborative approaches to achieve more-substantiated conclusions that all stakeholders can contribute to.

Here we review the formal body of literature on impact significance in EIA, with an emphasis on both technical and collaborative approaches to significance determination.

7.5.2 Integrating technical and collaborative approaches

The earliest frameworks for EIA (e.g., Leopold et al. 1971; Dee et al. 1973) relied on numerical weighting and scoring methods to assign importance values to potential development-environment interactions. Subsequent guidance materials for EIA (e.g., Sharma et al. 1976) sought to provide more rigorous definitions of the term 'significance' and to establish basic criteria (e.g., magnitude, extent, duration, frequency) for determining impact significance in EIA practice (e.g., Andrews et al. 1977). At the same time, members of the scientific community (e.g., Cairns 1977; Holling 1978) advocated concepts like assimilative capacity and stability domains when trying to gauge the ability of ecosystems to absorb or recover from anthropogenic disturbances. Still other authors (e.g., Bisset 1978) criticized early EIA frameworks for deliberately excluding diverse stakeholder perspectives from the evaluation of impact significance and, ultimately, from influencing development decisions.

Reviews of significance determination literature in the 1980s (e.g., Beanlands and Duinker 1983; Duinker and Beanlands 1986) outlined a range of technical, political, and regulatory perspectives on the subject. In general terms, Duinker and Beanlands (1986) highlighted some differences between significance judgements based on ecological limitations, limits of social acceptability, and regulatory environmental standards. According to Duinker and Beanlands (1986), a rigorous approach to significance determination should consider at minimum: (i) the social importance of the environmental attribute in question, (ii) the magnitude and distribution of predicted environmental impacts, and (iii) the reliability of impact predictions. From a scientific perspective, Conover et al. (1985) proposed a framework for evaluating impact significance based on the application of context-specific, ecological thresholds. At the same time, other authors proposed frameworks based on the application of more socially derived 'thresholds of concern' (e.g., Sassaman 1981; Haug et al. 1984).

More recently, the determination of impact significance has continued to be an important topic of discussion in the literature, with authors advocating a variety of tools and techniques. Examples include the use of so-called 'decision-trees' (e.g., Sippe 1999), 'fuzzy sets' (e.g., Silvert 1997; Wood et al. 2007), matrices (e.g., Ijäs et al. 2010; Toro et al. 2013), multi-criteria decision methods (e.g., Cloquell-Ballester et al. 2007), spatial indices (e.g., Antunes et al. 2001), regulatory environmental standards (e.g., Kjellerup 1999; Schmidt et al. 2008), socially derived thresholds of acceptability (e.g., Canter and Canty 1993; Ehrlich and Ross 2015), and ecologically based response thresholds (e.g., Duinker and Greig 2006; Duinker et al. 2013). Though the purpose of significance determination has largely been recaptured in terms of sustainable development (e.g., Gibson 2001), generic criteria such as magnitude, extent, frequency, and duration have seen continued use throughout the literature. Despite a growing body of literature on the subject, reviews of theory (e.g., Thompson 1990; Lawrence 2007b, 2007c; Jones and Morrison-Saunders 2016) and practice (e.g., Sadler 1996; Barnes et al. 2002; Wood 2008; Khadka et al. 2011; Briggs and Hudson 2013) have highlighted ongoing shortcomings in the evaluation of impact significance.

To reconcile technical and collaborative contributions to significance determination, Lawrence (2007a) has proposed a more integrated approach. While the technical approach aims to judge impact significance quantitatively based on scientific analysis and knowledge, the collaborative approach aims to do so qualitatively based on community knowledge and perspectives. A third approach, called reasoned argumentation, seeks to integrate quantitative and qualitative approaches. According to Lawrence (2007a), "The overall intent is to contribute to significance determination procedures that are less biased and distorted, more fully substantiated, more open, inclusive and collaborative and more effectively linked to decision making". In our view, the reasoned argumentation approach represents a much stronger framework for significance determination than either the technical or the collaborative approach used alone.

7.6 Evaluation of alternatives

7.6.1 Overview

Closely related to the evaluation of impact significance is the evaluation and selection of development alternatives. As pointed out by Wood (1995), decision-making in EIA may involve relatively simple yes/no regulatory approvals, or more complex decisions involving a choice among alternatives. According to Lawrence (1993), when development decisions involve multiple alternatives, environmental effects, and stakeholder perspectives, formal evaluation tools may be needed to handle a large number of trade-offs, but also to ensure transparency in the selection of a preferred alternative. Lawrence (1993) points out that there are two broad approaches to evaluating development alternatives in EIA: quantitative and qualitative. Therein, the quantitative approach is to evaluate alternatives numerically, relying primarily on abstract weighting, scaling, and aggregation techniques. The qualitative approach aims to evaluate development alternatives deliberatively, incorporating a range of stakeholder values and perspectives. Similar to the reasoned argumentation approach to impact significance determination (Lawrence 2007a), Lawrence (1993) proposed the integration of quantitative and qualitative approaches to the evaluation and selection of development alternatives. Such an approach, he argues, aims to contribute to more-substantiated and democratic environmental decisions.

In this section, we review the literature on evaluating development alternatives in EIA, with coverage of both quantitative and qualitative approaches.

7.6.2 *Integrating quantitative and qualitative approaches*

The earliest frameworks for EIA served as rapid evaluation tools for numerically weighting, aggregating, and comparing potential development-environment interactions. While early checklist/matrix approaches (e.g., Leopold et al. 1971; Dee et al. 1973) relied on the use of dimensionless impact scores to quantify development-environment interactions, cost-benefit analysis (e.g., Kasper 1977; Bohm and Henry 1979) required all environmental impacts to be translated into monetary terms, aggregated, and subtracted from benefits. Both cost-benefit analysis and early matrix/checklist methods received much criticism in the early literature (e.g., Lapping 1975; Pearce 1976). From both technical and regulatory perspectives, Andrews (1973) criticized such approaches for relying on the use of arbitrary, dimensionless impact scores or dollar amounts to quantify trade-offs. From a political perspective, Bisset (1978) criticized such approaches for being undemocratic and for concealing the basis upon which numerical values are assigned to impacts by experts.

Early scientific guidance materials for EIA (e.g., Holling 1978; Munn 1979) proposed somewhat more rigorous approaches to evaluating development alternatives based on the predicted future condition of actual measurable environmental attributes (i.e., indicators). Holling (1978) observed that when the number of development alternatives and indicators is large, it may be helpful to quantify the trade-off preferences of decision-makers and use techniques of mathematical optimization to determine the 'best' possible alternative. He also noted, however, that such complicated evaluation techniques may often be unnecessary or even counterproductive. Holling (1978) concluded that although quantitative evaluation techniques may be helpful when decision-makers are faced with several alternative developments and indicators, a simple visual examination of a side-by-side time-series or other graphical depiction of indicator patterns should generally suffice. According to Holling (1978), such a qualitative approach to evaluation helps to ensure that all parties are able to judge the predicted impacts of development alternatives from their own perspectives.

In the 1980s, the participatory dimension of EIA began to receive more critical attention in the literature (e.g., Rosenberg et al. 1981; Rossini and Porter 1982). These authors called for more-inclusive approaches to conducting EIAs and to making environmental decisions. In place of more-traditional approaches to making group decisions and settling disputes (i.e., quasi-judicial hearings, litigation), several consensus-building techniques were proposed for the evaluation and selection of development alternatives. Typical examples included the use of mediation (e.g., Cormick 1980; Sorensen et al. 1984) and the use of Delphi method (e.g., Richey et al. 1985; Miller and Cuff 1986). Still, with the rapid expansion of more powerful computers, many authors continued to advocate the use of quantitative aids for structuring complex decisions. Examples included so-called expert or decision support systems (e.g., Hushon 1987; Lein 1989), multi-criteria decision methods (e.g., Hobbs 1985; Nijkamp 1986), and optimal control theory (e.g., Walters 1986).

At the same time, several authors continued to highlight the inadequacies of both quantitative and qualitative approaches to evaluating development alternatives (e.g., McAllister 1980; Hollick 1981; Bakus et al. 1982). To move towards a more-inclusive, transparent, and less-biased approach to environmental decision-making, authors like Bakus et al. (1982) and Edwards and von Winterfeldt (1987) proposed frameworks that would use multiple-criteria decision methods as a basis for structuring stakeholder consensus-building and negotiation processes.

In many ways, the literature of the 1990s and 2000s has been a continuation of earlier literature on evaluation and decision-making in EIA. With respect to quantitative approaches, there has been an increasing focus on multi-criteria decision methods for evaluating alternatives (e.g., Janssen 2001; Huth et al. 2004; Kain and Söderberg 2008). At the same time, there has been continued criticism of traditional quantitative techniques like cost-benefit analysis (e.g., Munda 1996; Brunner and Starkl 2004).

With respect to qualitative approaches, several authors (e.g., Lee 1993; Gibson 2006) have continued to highlight the use of rule-based negotiation and mediation, while other authors (e.g., Kingston et al. 2000; Konisky and Beierle 2001) have proposed more innovative arrangements like citizen juries, study circles, roundtables, and web-based forums. Still, reviews of decision-making theory (e.g., Ortolano and Shepherd 1995; Pischke and Cashmore 2006) and practice (e.g., van Breda and Dijkema 1998; Morrison-Saunders and Bailey 2000; Leknes 2001; Richardson 2005) have highlighted ongoing weaknesses in the link between EIA and decision-making.

To enhance the influence of EIA on development planning and decision-making, many authors (e.g., Sinclair and Diduck 1995; Shepherd and Bowler 1997; Petts 1999; Doelle and Sinclair 2006; Stewart and Sinclair 2007) have called for stronger participatory processes that encourage early and ongoing stakeholder participation, particularly with respect to the design, evaluation, and selection of development alternatives.

To reconcile differing perspectives on the subject, Lawrence (1993) proposed the integration of qualitative and quantitative approaches to evaluation and decision-making in EIA. Such an integrated approach, he argued, would provide more substantiated and democratic conclusions than either of the two approaches used alone. Indeed, many authors have proposed the use of multi-criteria decision analysis to support more-structured and more-deliberative group decision-making processes (e.g., Ramanathan 2001; Petts 2003; Mardle et al. 2004; Kalbar et al. 2013). In our view, combining such quantitative and qualitative approaches to evaluation may be helpful for achieving mutually agreeable solutions when decision-makers are faced with several development alternatives, environmental impacts, and stakeholder objectives.

7.7 Formal reviews

7.7.1 Overview

Founded on principles of public scrutiny, EIA processes generally allow for public review of EIA documents prior to regulatory decision-making. In many jurisdictions, environmental laws and regulations require EIAs for large and controversial undertakings to be reviewed by independent panels (e.g., Wood 1995; Sadler 1996; Ross 2000). Such panels typically evaluate an EIA's overall quality against established scoping guidelines or terms of reference, and then submit a set of recommendations to regulatory decision-makers. To inform such recommendations, review panels typically hold formal review hearings centred on the evaluation of a draft EIA report. Here public comments are invited, expert testimony is received, and additional information is requested from proponents. At the end of the review process, the proponent publishes a final EIA report and the review panel submits its final recommendations to regulatory decision-makers.

In this section we briefly summarize the literature on formal EIA document reviews, focusing on the intersection of administrative, scientific, and political dimensions.

7.7.2 Administrative, scientific, and political dimensions

The first EIA processes in North America relied primarily on administrative and judicial mechanisms for review (e.g., Leventhal 1974). According to the literature (e.g., Lynch 1972; Karp 1978), such reviews were often the source of much confusion and frustration among EIA participants due to a lack of general standards for judging the adequacy of draft EIAs. This typically resulted in prolonged and costly disagreement among EIA participants. The sprawling and descriptive nature of early EIA reports was often attributed to this threat of judicial review (e.g., Wichelman 1976). Such arguments postulated that proponents might intentionally prepare encyclopedic EIAs, thereby giving the impression of comprehensive coverage while simultaneously obscuring relevant or controversial environmental impacts. Indeed, the prevailing yet unrealistic expectation of EIA administrators and reviewers for complete environmental information was also recognized in the literature (e.g., Andrews 1973; Carpenter 1976).

To encourage early agreement among all EIA participants on the design, preparation, and review of assessments, EIA administrators in North America introduced the first formal requirements for scoping in the late 1970s (e.g., Fisher 1979; Hourcle 1979). Such procedures were intended to foster a more focused and collaborative approach to designing, conducting, and inevitably reviewing the quality of EIAs.

In the early 1980s, some authors (e.g., Beanlands and Duinker 1983; Carpenter 1983) lamented the costly, reactive, and generally frustrating nature of formal EIA reviews. Carpenter (1983) and Beanlands and Duinker (1983) observed that the sprawling character of formal scoping guidelines seemed to perpetuate many of the problems formerly attributed to adversarial review hearings. The lingering issue, they argued, was the failure of EIA practitioners and members of the scientific community to come together and agree on a common basis for designing, conducting, and reviewing scientific contributions to EIA. In other words, there was still no widespread agreement – even amongst scientists themselves – as to what could reasonably be accomplished by science in EIA. What they could agree on, however, was the need for greater focus and the need for basic scientific standards. In the Canadian context, Beanlands and Duinker (1983) outlined six so-called requirements that were to serve as basic principles for both the design and evaluation of scientific contributions to EIA.

Later, Ross (1987) outlined the roles of different evaluators in judging the three major aspects of an EIA report: (i) focus, (ii) scientific and technical soundness, and (iii) clarity. According to Ross (1987), it should be the role of scientific and technical experts to judge matters of science, the role of the public to evaluate matters of focus and clarity, the role of administrators to test EIAs against scoping guidelines, and the role of independent panels to synthesize all comments and contributions into a set of recommendations. In sum, an ideal EIA should be focused on stakeholder-relevant issues, scientifically sound, and yet readily understandable to technical specialists and lay stakeholders alike.

According to Wood (1995), "If there is only one point in an EIA process where formal consultation and participation take place it is during the review of the EIA report". He goes to explain how in many jurisdictions, public review is virtually synonymous with public participation. According to several authors (e.g., Diduck and Sinclair 2002; Petts 2003; Doelle and Sinclair 2006; Stewart and Sinclair 2007), such formal mechanisms have generally failed to provide citizens with an opportunity to influence development decisions, and have thus failed to satisfy basic principles of participatory democracy. They argue that formal procedures for citizen engagement may even represent barriers to participatory environmental decision-making. Shepherd and Bowler (1997) argue that in order to improve the quality of participatory practices in EIA, proponents must go beyond formal requirements simply to consult the public before and after an EIA is prepared. Furthermore, they argue that it is often proponents who stand to benefit the most from inviting all stakeholders to participate in the design, evaluation, and selection of a preferred development alternative.

Still, the formal review of draft EIAs as a means of quality control continues to be discussed in the literature (e.g., Lee and Brown 1992; Ross 2000; Ross et al. 2006). More recently, the so-called review package outlined by Lee and Colley (1991) has become a popular method for reviewing and rating the quality of EIA documents (e.g., Cashmore et al. 2002; Canelas et al. 2005). Põder and Lukki (2011), however, have criticized such methods for emphasizing completeness of information over actual information quality. According to Ross et al. (2006), "The principle for preparing an EIS is simple; it should present a clear, concise summary of the likely environmental impacts, the proposed mitigation measures, the significance of the residual impacts and suggestions for needed follow-up studies".

7.8 Follow-up

7.8.1 Overview

According to Morrison-Saunders and Arts (2004), "A key feature of EIA is that its deals with the future and consequently is intrinsically uncertain. Follow-up addresses the uncertainties in EIA". Arts et al. (2001) identify four major activities within EIA follow-up: (i) monitoring, (ii) evaluation, (iii) management, and (iv) communication. Therein, monitoring is the ongoing collection of data, evaluation is the appraisal of those data's conformance with expectations, management is the set of actions taken in response to those evaluations, and communication is the disclosure of results to stakeholders. Arts et al. (2001) outline five major objectives for EIA follow-up: (i) provide information about the consequences of an activity, (ii) enhance scientific knowledge about environmental systems and cause-effect relationships, (iii) improve the quality of assessment methods and techniques, (iv) improve public awareness about the effects of development, and (v) maintain decision-making flexibility.

In this section, we provide two short reviews: one of the general literature on EIA follow-up and another focusing more specifically on the scientific literature surrounding environmental effects monitoring.

7.8.2 Effects monitoring, adaptive management, and participation

Holling (1978) argued that although accurate prediction of environmental impacts may be fundamentally impossible, some 'postdiction' (i.e., monitoring) might contribute to better predictions in the future. Holling (1978) further argued that effects monitoring would provide an important opportunity to pursue model invalidation, thereby reducing key uncertainties in the long run. To ensure the responsiveness of developments to new ecological information, Holling (1978) proposed an integrated brand of management and science (i.e., adaptive management). In Holling's (1978) adaptive management framework, not only are modelling and impact prediction integrated with development planning and design, but also with long-term ecological monitoring and management. Holling (1978) also pointed out the difference between regulatory monitoring, aimed at ensuring compliance with development approval conditions (e.g., mitigation measures), and scientific monitoring, aimed at testing impact predictions to reduce key uncertainties.

One of the six so-called requirements outlined by Beanlands and Duinker (1983) was to monitor the effects of developments on VECs.

Such monitoring efforts would be aimed at testing impact predictions, thus improving ecological effects knowledge for future EIAs. One aspect of effects monitoring to receive particularly close attention in the 1980s was the statistical design of field sampling programs, as well as the challenges of separating development-induced changes from natural variation. Whereas some authors (e.g., Green 1989) advocated simple before-after or control-impact comparisons to define an environmental impact, other authors (e.g., Stewart-Oaten et al. 1986) advocated a so-called before-after-control-impact (BACI) design. According to Stewart-Oaten et al.'s (1986) BACI framework, an environmental impact is defined as a change in the difference between measured environmental conditions at a control site and a treatment site following development. Duinker (1989) later argued that the establishment of such spatial and temporal controls is often impossible, particularly when dealing with highly mobile animal populations. Duinker (1989) reasoned that an environmental impact can only realistically be defined as the difference between the predicted future condition of an environmental component of interest with and without the proposed development in place. Consequently, Duinker (1989) concluded that monitoring can only realistically replace one of the predicted future time-series in a difference calculation of impact.

The participatory dimension of EIA follow-up has recently been a topic of considerable discussion in the literature (e.g., Lawe et al. 2005; Morrison-Saunders and Arts 2005). It is argued that enhancing stakeholder participation in EIA follow-up increases the utility of an assessment to all parties involved. According to Morrison-Saunders and Arts (2004), "Some follow-up programmes extend beyond simple communication to specifically include direct stakeholder participation in the monitoring, evaluation, and management steps as well". The literature describes a range of community-based (e.g., Lawe et al. 2005; O'Faircheallaigh 2007) and proponent-led (e.g., Marshall 2005; Noble and Storey 2005; Noble and Birk 2011) EIA follow-up programs. According to advocates of a community-led approach, such arrangements offer a level of transparency and trust not possible under proponent-driven initiatives. Most authors now agree that involving all stakeholders in follow-up processes allows EIA to have a greater influence on development outcomes as well as the long-term adaptive management of environmental effects. Indeed, established principles for good follow-up (e.g., Arts et al. 2001; Morrison-Saunders and Arts 2004; Marshall et al. 2005) emphasize regulatory, scientific, and participatory dimensions of EIA.

From a scientific perspective, several papers on environmental effects monitoring (e.g., Adams 2003; Hewitt et al. 2003) have recently focused on the need to explicitly link measured environmental effects with particular causes (i.e., stressors), thus reducing uncertainty in cause-effect relationships. It is generally argued that such approaches to causal inference are most useful when dealing with numerous, confounding, or otherwise uncertain causal variables (e.g., complex mill effluents). Several authors (e.g., Roux et al. 1999; Dubé and Munkittrick 2001; Kilgour et al. 2007) have recently proposed frameworks that integrate so-called stressor-based and response-based approaches to environmental effects monitoring. According to Roux et al. (1999), the stressor-based approach focuses on setting and enforcing regulatory standards for controlling levels or concentrations of particular stressors. Conversely, the response-based approach is focused on measuring ecological conditions in the presence of development. According to Roux et al. (1999), integrated stressor-response monitoring simultaneously measures both stressor and effect variables during development, thus providing scientific evidence to inform the adaptive management of stressors.

Several other authors (e.g., Lowell et al. 2000; Adams 2003) have recently outlined protocols for a so-called weight-of-evidence approach to inferring causality in environmental effects monitoring. Adams (2003) outlines seven causal criteria for evaluating a potential relationship between a stressor and an observed effect: (i) strength of association between stressor and effect variables, (ii) consistency of association between stressor and effect variables, (iii) specificity of association between stressor and effect variables, (iv) time order of stressor and effect measurement, (v) spatial or temporal dose-response gradient, (vi) experimental evidence, and (vii) a scientifically plausible explanation for a mechanism linking the proposed cause and effect.

According to Greig and Duinker (2011), it is the role of science inside EIA to predict and monitor the environmental effects of developments, thereby testing and refining the predictive tools developed by science outside EIA. In this way, environmental effects monitoring provides the critical feedback needed by science outside EIA to contribute better cause-effect knowledge in the future. Though the scientific literature reveals a growing body of knowledge surrounding the environmental effects of human developments, several authors (e.g., Karkkainen 2008; Bjorkland 2013; Roach and Walker 2017) observe ongoing inadequacies in how environmental effects monitoring is practiced in EIA. In this last section, we provide a general outline of the environmental effects knowledge published in the formal scientific literature. While our review highlights several studies that conceptually resemble environmental effects monitoring in EIA, a few examples from formal EIA practice have also been included.

7.8.3 Environmental effects knowledge

Most empirical studies of development-induced environmental effects conducted in the 1970s were in the form of brief comparisons of conditions measured at development and control sites. Less common were the results of actual long-term monitoring programs. Examples included studies on the effects of mining and wastewater discharge on marine and freshwater quality (e.g., Balch et al. 1976; Kaufmann et al. 1976), the effects of hydroelectric and thermal energy generation on freshwater fish (e.g., Merriman and Thorpe 1976; Mathur et al. 1977), the effects of pulp mill effluents on freshwater fish (e.g., Kelso 1977; Leslie and Kelso 1977), the effects of road-kill and hunting on wildlife populations (e.g., McCaffery 1973), and the effects of industrial food and timber production on soil nutrient budgets (e.g., Brown et al. 1973; Olness et al. 1975).

At the strategic level, the 1970s saw the initiation of some of the first monitoring programs aimed at characterizing the broad-scale environmental consequences of persistent air and water pollution. Examples included initiatives for monitoring the accumulation of pollutants like PCBs and methylmercury in freshwater and marine fish (e.g., Munson and Huggett 1972; Walker 1976; Hattula et al. 1978) and programs for monitoring the effects of acid deposition on water, forests, and fish (e.g., Ottar 1976; Galloway et al. 1978).

In the 1980s, empirical research on environmental impacts expanded to include the effects of habitat loss and landscape fragmentation on wildlife. Early examples included studies on the effects of roads, buildings, and other noisy infrastructure on wildlife habitat use (e.g., Witmer and deCalesta 1985; Mattson et al. 1987a; McLellan and Shackleton 1988) and the effects of timber harvesting on the abundance and diversity of birds (e.g., Scott and Oldemeyer 1983; Zarnowitz and Manuwal 1985). Still other studies continued to document the effects of hydroelectric and thermal energy developments on fish (e.g., Barnthouse et al. 1983; Bodaly et al. 1984), the effects of pulp mill effluents on fish (e.g., Andersson et al. 1988; Sandström and Thoresson 1988), the effects of mining on surface and groundwater quality (e.g., Hill and Price 1983), and the effects of timber harvesting on watershed nutrient and water budgets (e.g., Krause 1982; Feller and Kimmins 1984).

The 1980s also saw growing interest in monitoring the environmental effects of fish farming, an emerging industry with uncertain and potentially unwanted environmental consequences. Early examples included studies on the effects of caged fish farming on marine water and sediment quality (e.g., Kaspar et al. 1988; Eng et al. 1989; Frid and Mercer 1989).

With the expansion of GIS and satellite remote sensing, the 1980s would eventually see the use of such platforms to monitor the cumulative effects of developments on coastal wetlands (e.g., Nayak et al. 1989). Satellite observation systems would also be used to monitor the effects of deforestation and land-use change on the productivity of tropical forests around the world (e.g., Malingreau et al. 1989).

More recently, there has been an increase in published scientific research on environmental effects. Whereas most empirical effects research published in the 1970s and 1980s was in the form of short-term before-after or control-impact studies, more recent papers (e.g., Roux et al. 1999; Dubé and Munkittrick 2001; Adams 2003, Hewitt et al. 2003; Kilgour et al. 2007) have highlighted the importance of measuring both stressor and response variables at the same time to reduce uncertainty in cause-effect relationships. Noteworthy examples of such an approach include studies on the effects of pulp mill effluent on water and fish (e.g., Culp et al. 2000; Munkittrick et al. 2002; Walker et al. 2002; Hewitt et al. 2005; Dubé et al. 2006; Squires et al. 2010), the effects of mining effluent on water and fish (e.g., Ribey et al. 2002), and the effects of thermal energy developments on fish (e.g., Barnthouse 2000).

The recent literature also highlights many studies that have used biotelemetry and satellite remote sensing to characterize the effects of linear developments (e.g., roads) on wildlife-habitat relationships. Examples include studies on the effects of forest roads and cut-blocks on habitat use by wolves (e.g., Whittington et al. 2005; Houle et al. 2010) and studies on the effects of roads, pipelines, and seismic lines on habitat use by caribou (e.g., Dyer et al. 2002; Johnson et al. 2005; Sorensen et al. 2008; Polfus et al. 2011).

Emerging industries and technologies – surrounded by considerable uncertainty with respect to environmental impacts – have also been the subject of empirical scientific research. Examples include studies on the effects of wind energy turbines on birds and bats (e.g., Johnson et al. 2003; Plonczkier and Simms 2012; Parisé and Walker 2017), the effects of aquaculture on marine and freshwater quality (e.g., Boaventura et al. 1997; Selong and Helfrich 1998; La Rosa et al. 2002), and the effects of shale-gas fracking on surface and groundwater quality (e.g., Olmstead et al. 2013; Warner et al. 2013).

With growing concerns over the potential consequences of global climate change, the recent literature has also highlighted many initiatives for monitoring the effects of development on terrestrial carbon stocks. At the project level, such studies may monitor the effects of timber harvesting and forest management on local soil and tree carbon stocks (e.g., Prescott et al. 2000; Finkral and Evans 2008). At the strategic level, such initiatives may monitor the effects of land-use change on national soil and tree carbon stocks (e.g., Scott et al. 2002; Tate et al. 2003; Kurz and Apps 2006).

According to Greig and Duinker (2011), "Science inside the EIA process is needed to make specific impact predictions to inform decision-makers of the potential ecological consequences of development alternatives, as well as to measure environmental responses following development start-up for the purpose of model evaluation and refinement". With respect to monitoring, Greig and Duinker (2011) conclude that "Regardless of whether the assignment falls to scientists outside or practitioners inside the EIA process, the monitoring must get done or else reliable knowledge will not, over time, be developed". To this they add that "On the practical side, EIA should be able to experience gradually reduced costs with implementation of strong science in reducing impact uncertainty".

References

Aber, J.D., Botkin, D.B., and Melillo, J.M. 1978. Predicting the effects of different harvesting regimes on forest floor dynamics in northern hardwoods. *Canadian Journal of Forest Research* 8(3): 306–315.

Abrams, M.D., and Hulbert, L.C. 1987. Effect of topographic position and fire on species composition in tall grass prairie in northeast Kansas. *American Midland Naturalist* 117(2): 442–445.

Adams, M.B. 2011. Land application of hydro fracturing fluids damages a deciduous forest stand in West Virginia. *Journal of Environmental Quality* 40(4): 1340–1344.

Adams, S.M. 2003. Establishing causality between environmental stressors and effects on aquatic ecosystems. *Human and Ecological Risk Assessment* 9(1): 17–35.

Akçakaya, H.R. 2001. Linking population-level risk assessment with landscape and habitat models. *Science of the Total Environment* 274(1–3): 283–291.

Anderson, L., Carlson, C.E., and Wakimoto, R.H. 1987. Forest fire frequency and western spruce budworm outbreaks in western Montana. *Forest Ecology and Management* 22(3): 251–260.

Andersson, T., Förlin, L., Härdig, J., and Larsson, Å. 1988. Physiological disturbances in fish living in coastal water polluted with bleached kraft pulp mill effluents. *Canadian Journal of Fisheries and Aquatic Sciences* 45(9): 1525–1536.

Andrews, R.N.L. 1973. A philosophy of environmental impact assessment. *Journal of Soil and Water Conservation* 28: 197–203.

Andrews, R.N.L., Cromwell, P., Enk, G.A., Farnworth, E.G., Hibbs, J.R., and Sharp, V.L. 1977. *Substantive Guidance for Environmental Impact Assessment: An Exploratory Study*. Washington, DC: Institute of Ecology. 79 p.

Antunes, P., Santos, R., and Jordão, L. 2001. The application of geographical information systems to determine environmental impact significance. *Environmental Impact Assessment Review* 21(6): 511–535.

Armstrong, J.S. 1999. Forecasting for environmental decision making. In: *Tools to Aid Environmental Decision Making*. Dale, V.H. and English, M.R., editors. New York: Springer. 192–225 p.

Arts, J., Caldwell, P., and Morrison-Saunders, A. 2001. Environmental impact assessment follow-up: Good practice and future directions – findings from a workshop at the IAIA 2000 conference. *Impact Assessment and Project Appraisal* 19(3): 175–185.

Ausmus, B.S., Dodson, G.J., and Jackson, D.R. 1978. Behavior of heavy metals in forest microcosms. *Water Air & Soil Pollution* 10(1): 19–26.

Baker, W.L. 1989. A review of models of landscape change. *Landscape Ecology* 2(2): 111–133.

Bakus, G.J., Stillwell, W.G., Latter, S.M., and Wallerstein, M.C. 1982. Decision making: With applications for environmental management. *Environmental Management* 6(6): 493–504.

Balch, N., Ellis, D., Littlepage, J., Marles, E., and Pym, R. 1976. Monitoring a deep marine wastewater outfall. *Journal of Water Pollution Control Federation* 48(3): 429–457.

Barnes, J.L., Matthews, L., Griffiths, A., and Horvath, C.L. 2002. Addressing cumulative environmental effects: determining significance. In: *Cumulative Environmental Effects Management: Tools and Approaches*. Kennedy, A.J., editor. Edmonton: Alberta Society of Professional Biologists. 327–343 p.

Barnthouse, L.W. 2000. Impacts of power-plant cooling systems on estuarine fish populations: The Hudson River after 25 years. *Environmental Science & Policy* 3(1): S341-S348.

Barnthouse, L.W., Van Winkle, W., and Vaughan, D.S. 1983. Impingement losses of white perch at Hudson River power plants: Magnitude and biological significance. *Environmental Management* 7(4): 355–364.

Baron, P.A., and Willeke, K., editors. 2001. *Aerosol Measurement: Principles, Techniques, and Applications*. Second ed. New York: John Wiley & Sons. 1131 p.

Bastos, R., Pinhanços, A., Santos, M., Fernandes, R.F., Vicente, J.R., Morinha, F., Honrado, J.P., Travassos, P., Barros, P., and Cabral, J.A. 2016. Evaluating the regional cumulative impact of wind farms on birds: How can spatially explicit dynamic modeling improve impact assessments and monitoring? *Journal of Applied Ecology* 53(5): 1330–1340.

Bauerle, B., Spencer, D.L., and Wheeler, W. 1975. The use of snakes as a pollution indicator species. *Copeia* 2: 366–368.

Bayley, S.E., Vitt, D.H., Newbury, R.W., Beaty, K.G., Behr, R., and Miller, C. 1987. Experimental acidification of a sphagnum-dominated peatland: First year results. *Canadian Journal of Fisheries and Aquatic Sciences* 44: 194–205.

Beanlands, G. 1988. Scoping methods and baseline studies in EIA. In: *Environmental Impact Assessment: Theory and Practice*. Wathern, P., editor. London: Routledge. 33–46 p.

Beanlands, G.E., and Duinker, P.N. 1983. *An Ecological Framework for Environmental Impact Assessment in Canada*. Halifax: Institute for Resource and Environmental Studies, Dalhousie University. 132 p.

Beanlands, G.E., Erckmann, W.J., Orians, G.H., O'Riordan, J., Policansky, D., Sadar, M.H., and Sadler, B. 1985. *Cumulative Environmental Effects: A Binational Perspective*. Proceedings of a Workshop. Hull: Canadian Environmental Assessment Research Council. 175 p.

Best, P.B., Brandao, A., and Butterworth, D.S. 2001. Demographic parameters of southern right whales off South Africa. *Journal of Cetacean Research and Management* 2: 161–169.

Bierregaard, R.O., and Lovejoy, T.E. 1989. Effects of forest fragmentation on Amazonian understory bird communities. *Acta Amazonica* 19: 215–41.

Billen, G., and Garnier, J. 1999. Nitrogen transfers through the Seine drainage network: A budget based on the application of the 'Riverstrahler' model. *Hydrobiologia* 410(0): 139–150.

Birch, A.N.E., Griffiths, B.S., Caul, S., Thompson, J., Heckmann, L.H., Krogh, P.H., and Cortet, J. 2007. The role of laboratory, glasshouse and field scale experiments in understanding the interactions between genetically modified crops and soil ecosystems: A review of the ECOGEN project. *Pedobiologia* 51(3): 251–260.

Bisset, R. 1978. Quantification, decision-making and environmental impact assessment in the United Kingdom. *Journal of Environmental Management* 7: 43–58.

Bjorkland, R. 2013. Monitoring: the missing piece: A critique of NEPA monitoring. *Environmental Impact Assessment Review* 43: 129–134.

Blaikley, D.R., Dietzel, G.F.L., Glass, A.W., and van Kleef, P.J. 1977. "SLIKTRAK" – a computer simulation of offshore oil spills, cleanup, effects and associated costs. In: *International Oil Spill Conference Proceedings*. Washington, DC: American Petroleum Institute. 45–52 p.

Boaventura, R., Pedro, A.M., Coimbra, J., and Lencastre, E. 1997. Trout farm effluents: characterization and impact on the receiving streams. *Environmental Pollution* 95(3): 379–387.

Bodaly, R.A., Hecky, R.E., and Fudge, R.J.P. 1984. Increases in fish mercury levels in lakes flooded by the Churchill River diversion, northern Manitoba. *Canadian Journal of Fisheries and Aquatic Sciences* 41(4): 682–691.

Bohm, P., and Henry, C. 1979. Cost-benefit analysis and environmental effects. *Ambio* 8(1): 18–24.

Bond, A.J., Morrison-Saunders, A., Gunn, J.A.E., Pope, J., and Retief, F. 2015. Managing uncertainty, ambiguity and ignorance in impact assessment by embedding evolutionary resilience, participatory modeling and adaptive management. *Journal of Environmental Management* 151: 97–104.

Booty, W.G., and Kramer, J.R. 1984. Sensitivity analysis of a watershed acidification model. *Philosophical Transactions of the Royal Society B* 305(1124): 441–449.

Botkin, D.B., Janak, J.F., and Wallis, J.R. 1972. Some ecological consequences of a computer model of forest growth. *Journal of Ecology* 60(3): 849–872.

Bradshaw, C., Veale, L.O., Hill, A.S., and Brand, A.R. 2001. The effect of scallop dredging on Irish Sea benthos: Experiments using a closed area. *Hydrobiologia* 465(1): 129–138.

Bridge, J.S., and Leeder, M.R. 1979. A simulation model of alluvial stratigraphy. *Sedimentology* 26(5): 617–644.

Briggs, S., and Hudson, M.D. 2013. Determination of significance in ecological impact assessment: Past change, current practice and future improvements. *Environmental Impact Assessment Review* 38: 16–25.

Brown, A.L., and Hill, R.C. 1995. Decision-scoping: Making EA learn how the design process works. *Project Appraisal* 10(4): 223–232.

Brown, G.W., Gahler, A.R., and Marston, R.B. 1973. Nutrient losses after clear-cut logging and slash burning in the Oregon Coast Range. *Water Resources Research* 9(5): 1450–1453.

Brown, S. 2002. Measuring carbon in forests: Current status and future challenges. *Environmental Pollution* 116(3): 363–372.

Brunner, N., and Starkl, M. 2004. Decision aid systems for evaluating sustainability: A critical survey. *Environmental Impact Assessment Review* 24(4): 441–469.

Buckland, S.T., Rexstad, E.A., Marques, T.A., and Oedekoven, C.S. 2015. *Distance Sampling: Methods and Applications*. Cham: Springer. 277 p.

Bunnell, F.L., Dunbar, D., Koza, L., and Ryder, G. 1981. Effects of disturbance on the productivity and numbers of white pelicans in British Columbia: Observations and models. *Colon Water Bird* 4: 2–11.

Cairns, J. 1974. Indicator species vs. The concept of community structure as an index of pollution. *Journal of the American Water Resource Association* 10(2): 338–347.

Cairns, J. 1977. Aquatic ecosystem assimilative capacity. *Fisheries* 2(2): 5–7.

Cairns, J. 1980. *The Recovery Process in Damaged Ecosystems*. Ann Arbor: Ann Arbor Science Publishers. 167 p.

Caldwell, L.K., Bartlett, R.V., Parker, D.E., and Keys, D.L. 1982. *A Study of Ways to Improve the Scientific Content and Methodology of Environmental Impact Analysis*. Bloomington: School of Public and Environmental Affairs, Indiana University. 453 p.

Callahan, J.T. 1984. Long-term ecological research. *Bioscience* 34(6): 363–367.

Canale, R.P., Patterson, R.L., Gannon, J.J., and Powers, W.F. 1973. Water quality models for total coliform. *Journal of Water Pollution Control Federation* 45(2): 325–336.

Canelas, L., Almansa, P., Merchan, M., and Cifuentes, P. 2005. Quality of environmental impact statements in Portugal and Spain. *Environmental Impact Assessment Review* 25(3): 217–225.

Canter, L.W., and Atkinson, S.F. 2011. Multiple uses of indicators and indices in cumulative effects assessment and management. *Environmental Impact Assessment Review* 31(5): 491–501.

Canter, L.W., and Canty, G.A. 1993. Impact significance determination – basic considerations and a sequenced approach. *Environmental Impact Assessment Review* 13(5): 275–297.

Cardenas, I.C., and Halman, J.I.M. 2016. Coping with uncertainty in environmental impact assessments: Open techniques. *Environmental Impact Assessment Review* 60: 24–39.

Carpenter, R.A. 1976. The scientific basis of NEPA – is it adequate? *Environmental Law Reporter* 6: 50014–50019.

Carpenter, R.A. 1983. Ecology in court, and other disappointments of environmental science and environmental law. *Natural Resources Law* 15: 573–596.

Carpenter, S.R., and Kitchell, J.F. 1988. Consumer control of lake productivity. *Bioscience* 38(11): 764–769.

Carpenter, S.R., Leavitt, P.R., Elser, J.J., and Elser, M.M. 1988. Chlorophyll budgets: Response to food web manipulation. *Biogeochemistry* 6(2): 79–90.

Science in the EIA process 107

Carpenter, S.B., Montgomery, T.L., Leavitt, J.M., Colbaugh, W.C., and Thomas, F.W. 1971. Principal plume dispersion models: TVA power plants. *Journal of Air Pollution Control Association* 21(8): 491–495.

Carroll, B.K., and Brown, D.L. 1977. Factors affecting neonatal fawn survival in southern-central Texas. *Journal of Wildlife Management* 41(1): 63–69.

Carter, M.R., and Gregorich, E.G., editors. 2008. *Soil Sampling and Methods of Analysis*. Second ed. Boca Raton: CRC Press. 1224 p.

Cashmore, M. 2004. The role of science in environmental impact assessment: Process and procedure versus purpose in the development of theory. *Environmental Impact Assessment Review* 24(4): 403–426.

Cashmore, M., Christophilopoulos, E., and Cobb, D. 2002. An evaluation of the quality of environmental impact statements in Thessaloniki, Greece. *Journal of Environmental Assessment Policy and Management* 4(4): 371–395.

CEQ (Council on Environmental Quality). 1974. *The Role of Ecology in the Federal Government*. Washington, DC: Council on Environmental Quality. 78 p.

CEQ (Council on Environmental Quality). 1978. National Environmental Policy Act, implementation of procedural provisions, final regulations. *Federal Register* 43: 55978–56007.

CEQ (Council on Environmental Quality). 1980. *Environmental Quality – the Eleventh Annual Report of the Council on Environmental Quality*. Washington, DC: Council on Environmental Quality. 497 p.

Chow, J.C. 1995. Measurement methods to determine compliance with ambient air quality standards for suspended particles. *Journal of the Air & Waste Management Association* 45(5): 320–382.

Christensen, V., and Walters, C.J. 2004. Ecopath with Ecosim: Methods, capabilities and limitations. *Ecological Modeling* 172(2–4): 109–139.

Clark, B.D., Bisset, R., and Tomlinson, P. 1983. *Post-Development Audits to Test the Effectiveness of Environmental Impact Prediction Methods and Techniques*. Aberdeen: PADC Environmental Impact Assessment and Planning Unit, University of Aberdeen. 47 p.

Clark, J.S., Carpenter, S.R., Barber, M., Collins, S., Dobson, A., Foley, J.A., Lodge, D.M., Pascual, M., Pielke, R., Pizer, W., et al. 2001. Ecological forecasts: An emerging imperative. *Science* 293(5530): 657–660.

Clark, R. 1994. Cumulative effects assessment: A tool for sustainable development. *Impact Assess* 12(3): 319–331.

Cloquell-Ballester, V., Monterde-Díaz, R., Cloquell-Ballester, V., and Santamarina-Siurana, M. 2007. Systematic comparative and sensitivity analyses of additive and outranking techniques for supporting impact significance assessments. *Environmental Impact Assessment Review* 27(1): 62–83.

Collins, S.L. 1987. Interaction of disturbances in tall grass prairie: A field experiment. *Ecology* 68(5): 1243–1250.

Collins, S.L., Bettencourt, L.M., Hagberg, A., Brown, R.F., Moore, D.I., Bonito, G., Delin, K.A., Jackson, S.P., Johnson, D.W., Burleigh, S.C., et al. 2006. New opportunities in ecological sensing using wireless sensor networks. *Frontiers in Ecology and the Environment* 4(8): 402–407.

Conover, S., Strong, K., Hickey, T., and Sander, F. 1985. An evolving framework for environmental impact analysis. I: Methods. *Journal of Environmental Economics and Management* 21: 343–358.

Cook, R.S., White, M., Trainer, D.O., and Glazener, W.C. 1971. Mortality of young white-tailed deer fawns in south Texas. *Journal of Wildlife Management* 35(1):47–56.

Cooke, S.J., Hinch, S.G., Wikelski, M., Andrews, R.D., Kuchel, L.J., Wolcott, T.G., and Butler, P.J. 2004. Biotelemetry: a mechanistic approach to ecology. *Trends in Ecology and Evolution* 19(6): 334–343.

Cooke, S.J., Midwood, J.D., Thiem, J.D., Klimley, P., Lucas, M.C., Thorstad, E.B., Eiler, J., Holbrook, C., and Ebner, B.C. 2013. Tracking animals in freshwater with electronic tags: Past, present and future. *Animal Biotelemetry* 1(1): 5.

Cormick, G.W. 1980. The "theory" and practice of environmental mediation. *The Environmental Professional* 2(1): 24–33.

Cortet, J., Griffiths, B.S., Bohanec, M., Demsar, D., Andersen, M.N., Caul, S.E., Birch, A.N., Pernin, C., Tabone, E., de Vaufleury, A., et al. 2007. Evaluation of effects of transgenic Bt maize on microarthropods in a European multi-site experiment. *Pedobiologia* 51(3): 207–218.

Cosby, B.J., Hornberger, G.M., Galloway, J.N., and Wright, R.E. 1985. Time scales of catchment acidification: A quantitative model for estimating freshwater acidification. *Environmental Science & Technology* 19(12): 1144–1149.

Costanza, R., Sklar, F.H., White, M.L., and Day, Jr J.W. 1988. A dynamic spatial simulation model of land loss and marsh succession in coastal Louisiana. In: *Wetland Modeling: Developments in Environmental Modeling 12.* Mitsch, W.J., Straskraba, M., and Jørgensen, S.E., editors. Amsterdam: Elsevier. 99–114 p.

Craighead, F.C., and Craighead, J.J. 1972. Grizzly bear prehibernation and denning activities as determined by radiotracking. *Wildlife Monographs* 32: 3–35.

Culhane, P.J. 1987. The precision and accuracy of U.S. environmental impact statements. *Environmental Monitoring and Assessment* 8(3): 217–238.

Culp, J.M., Cash, K.J., and Wrona, F.J. 2000. Cumulative effects assessment for the Northern River Basins Study. *Journal of Aquatic Ecosystem Stress and Recovery* 8(1): 87–94.

Daniels, A., and Bach, W. 1976. Simulation of the environmental impact of an airport on the surrounding air quality. *Journal of the Air Pollution Control Association* 26(4): 339–344.

De Leeuw, J., Georgiadou, Y., Kerle, N., De Gier, A., Inoue, Y., Ferwerda, J., Smies, M., and Narantuya, D. 2010. The function of remote sensing in support of environmental policy. *Remote Sensing* 2(7): 1731–1750.

DeAngelis, D.L., Christensen, S.W., and Clark, A.G. 1977. Responses of a fish population model to young-of-the-year mortality. *Journal of Fisheries Research Board of Canada* 34(11): 2117–2123.

Dee, N., Baker, J., Drobny, N., Duke, K., Whitman, I., and Fahringer, D. 1973. An environmental evaluation system for water resource planning. *Water Resources Research* 9(3): 523–535.

Demyanov, V., Wood, S.N., and Kedwards, T.J. 2006. Improving ecological impact assessment by statistical data synthesis using process-based models. *Journal of the Royal Statistical Society Series C Applied Statistics* 55(1): 41–62.

Deriso, R.B. 1980. Harvesting strategies and parameter estimation for an age-structured model. *Canadian Journal of Fisheries and Aquatic Sciences* 37(2): 268–282.

Dickinson, R.E., and Henderson-Sellers, A. 1988. Modeling tropical deforestation: A study of GCM land-surface parametrizations. *Quarterly Journal of the Royal Meteorological Society* 114(480): 439–462.

Diduck, A., and Sinclair, A.J. 2002. Public involvement in environmental assessment: The case of the nonparticipant. *Environmental Management* 29(4): 578–588.

Doelle, M., and Sinclair, A.J. 2006. Time for a new approach to public participation in EA: Promoting cooperation and consensus for sustainability. *Environmental Impact Assessment Review* 26(2): 185–205.

Dolbeer, R.A., and Clark, W.R. 1975. Population ecology of snowshoe hares in the central Rocky Mountains. *Journal of Wildlife Management* 39(3): 535–549.

Dorn, P., Johnson, L., and Darby, C. 1979. The swimming performance of nine species of common California inshore fishes. *Transactions of the American Fisheries Society* 108(4): 366–372.

Dubé, M., Johnson, B., Dunn, G., Culp, J., Cash, K., Munkittrick, K., Wong, I., Hedley, K., Booty, W., Lam, D., et al. 2006. Development of a new approach to cumulative effects assessment: A Northern River ecosystem example. *Environmental Monitoring and Assessment* 113(1): 87–115.

Dubé, M., and Munkittrick, K. 2001. Integration of effects-based and stressor-based approaches into a holistic framework for cumulative effects assessment in aquatic ecosystems. *Human and Ecological Risk Assessment* 7(2): 247–258.

Duinker, P.N. 1989. Ecological effects monitoring in environmental impact assessment: What can it accomplish? *Environmental Management* 13(6): 797–805.

Duinker, P.N., and Baskerville, G.L. 1986. A systematic approach to forecasting in environmental impact assessment. *Journal of Environmental Management* 23: 271–290.

Duinker, P.N., and Beanlands, G.E. 1986. The significance of environmental impacts: An exploration of the concept. *Environmental Management* 10(1): 1–10.

Duinker, P.N., Burbidge, E.L., Boardley, S.R., and Greig, L.A. 2013. Scientific dimensions of cumulative effects assessment: Toward improvements in guidance for practice. *Environmental Review* 21(1): 40–52.

Duinker, P.N., and Greig, L.A. 2006. The impotence of cumulative effects assessment in Canada: Ailments and ideas for redeployment. *Environmental Management* 37(2): 153–161.

Duinker, P.N., and Greig, L.A. 2007. Scenario analysis in environmental impact assessment: Improving explorations of the future. *Environmental Impact Assessment Review* 27(3): 206–219.

Duncan, R. 2008. Problematic practice in integrated impact assessment: The role of consultants and predictive computer models in burying uncertainty. *Impact Assessment and Project Appraisal* 26(1): 53–66.

Dunning, J.B., Stewart, D.J., Danielson, B.J., Noon, B.R., Root, T.L., Lamberson, R.H., and Stevens, E.E. 1995. Spatially explicit population models: Current forms and future uses. *Ecological Applications* 5(1): 3–11.

Dutt, G.R., Shaffer, M.J., and Moore, W.J. 1972. *Computer Simulation Model of Dynamic Bio-Physico-Chemical Processes in Soils. Technical Bulletin 196.* Tucson: Agricultural Experimental Station, University of Arizona. 101 p.

Dutta, P., Mahatha, S., and De, P. 2004. A methodology for cumulative impact assessment of opencast mining projects with special reference to air quality assessment. *Impact Assessment and Project Appraisal* 22(3): 235–250.

Dyer, S.J., O'Neill, J.P., Wasel, S.M., and Boutin, S. 2002. Quantifying barrier effects of roads and seismic lines on movements of female woodland caribou in northeastern Alberta. *Canadian Journal of Zoology* 80(5): 839–845.

Edwards, W., and von Winterfeldt, D. 1987. Public values in risk debates. *Risk Analysis* 7(2): 141–158.

Ehrlich, A., and Ross, W. 2015. The significance spectrum and EIA significance determinations. *Impact Assessment and Project Appraisal* 33(2): 87–97.

Eichhorn, M., Johst, K., Seppelt, R., and Drechsler, M. 2012. Model-based estimation of collision risks of predatory birds with wind turbines. *Ecology and Society* 17(2): 1.

Ek, A.R., and Monserud, R.A. 1974. *FOREST: A Computer Model for the Growth and Reproduction of Mixed Species Forest Stands*. Research Report A2635. Madison: College of Agricultural and Life Sciences, University of Wisconsin-Madison. 90 p.

Eng, C.T., Paw, J.N., and Guarin, F.Y. 1989. The environmental impact of aquaculture and the effects of pollution on coastal aquaculture development in Southeast Asia. *Marine Pollution Bulletin* 20(7): 335–343.

Evans, E.C. 1977. Microcosm responses to environmental perturbants. *Helgolander Wissenschaftliche Meeresuntersuchungen* 30(1): 178–191.

Fancy, S.G., Frank, L.F., Douglas, D.C., Curby, C.H., Garner, G.W, Amstrup, S.C., and Regelin, W.L. 1988. *Satellite Telemetry: A New Tool for Wildlife Research and Management*. FWS-PUB-172. Washington, DC: US Department of the Interior. 54 p.

Fancy, S.G., Pank, L.F., Whitten, K.R., and Regelin, W.L. 1989. Seasonal movements of caribou in arctic Alaska as determined by satellite. *Canadian Journal of Zoology* 67(3): 644–650.

Fancy, S.G., Whitten, K.R., and Russell, D.E. 1994. Demography of the Porcupine caribou herd, 1983–1992. *Canadian Journal of Zoology* 72(5): 840–846.

Feller, M.C, and Kimmins, J.P. 1984. Effects of clear cutting and slash burning on stream water chemistry and watershed nutrient budgets in southwestern British Columbia. *Water Resources Research* 20(1): 29–40.

Ferrier, R.C., Whitehead, P.G., Sefton, C., Edwards, A.C., and Pugh, K. 1995. Modeling impacts of land use change and climate change on nitrate-nitrogen in the River Don, North East Scotland. *Water Resources* 29(8): 1950–1956.

Fetter, C.W. 2000. *Applied Hydrogeology*. Fourth ed. Upper Saddle River: Prentice Hall. 598 p.

Findlay, D.L., Podemski, C.L., and Kasian, S.E.M. 2009. Aquaculture impacts on the algal and bacterial communities in a small boreal forest lake. *Canadian Journal of Fisheries and Aquatic Sciences* 66(11): 1936–1948.

Finkral, A.J., and Evans, A.M. 2008. The effects of a thinning treatment on carbon stocks in a northern Arizona ponderosa pine forest. *Forest Ecology and Management* 255(7): 2743–2750.

Fisher, M. 1979. The CEQ regulations: New stage in the evolution of NEPA. *Harvard Environmental Law* 3: 347–380.

Forman, R.T., and Godron, M. 1981. Patches and structural components for a landscape ecology. *Bioscience* 31(10): 733–740.

Franklin, J. 2010. Moving beyond static species distribution models in support of conservation biogeography. *Diversity and Distributions* 16(3): 321–330.

Freddy, D., Carpenter, L., Spraker, T., Strong, L., Neil, P., McCloskey, B., Sweeting, J., Ward, L., and Cupal, J. 1977. *Snowmobile Harassment of Mule Deer on Cold Winter Ranges*. Job Progress Report, Project No. W-38-R-32. Denver: US Forest Service. 15 p.

Freeman, M.M.R. 1992. The nature and utility of traditional ecological knowledge. *Northern Perspectives* 20(1): 9–12.

Freeze, R.A. 1971. Three-dimensional, transient, saturated-unsaturated flow in a groundwater basin. *Water Resources Research* 7(2): 347–366.

Frid, C.L.J., and Mercer, T.S. 1989. Environmental monitoring of caged fish farming in macrotidal environments. *Marine Pollution Bulletin* 20(8): 379–383.

Friedl, M.A., McIver, D.K., Hodges, J.C.F., Zhang, X.Y., Muchoney, D., Strahler, A.H., Woodcock, C.E., Gopal, S., Schneider, A., Cooper, A., et al. 2002. Global land cover mapping from MODIS: Algorithms and early results. *Remote Sensing of Environment* 83(1–2): 287–302.

Gagnon, C.A., and Berteaux, D. 2009. Integrating traditional ecological knowledge and ecological science: A question of scale. *Ecology and Society* 14(2): 19.

Galloway, J.N., Cowling, E.B., Gorham, E., and McFee, W.M. 1978. *A National Program for Assessing the Problem of Atmospheric Deposition (Acid Rain)*. A Report to the Council on Environmental Quality. Fort Collins: National Atmospheric Deposition Program. 97 p.

Gelhar, L.W., and Wilson, J.L. 1974. Ground-water quality modeling. *Ground Water* 12(6): 399–408.

Geneletti, D. 2006. Some common shortcomings in the treatment of impacts of linear infrastructures on natural habitat. *Environmental Impact Assessment Review* 26(3): 257–267.

Gibbs, H.K., Brown, S., O Niles, J., and Foley, J.A. 2007. Monitoring and estimating tropical forest carbon stocks: Making REDD a reality. *Environmental Research Letters* 2(4): 045023.

Gibson, D.J. 1989. Hulbert's study of factors affecting botanical composition of tallgrass prairie. In: *Proceedings of the Eleventh North American Prairie Conference*. Bragg, T. and Stubbendieck, J., editors. Lincoln: University of Nebraska. 115–133 p.

Gibson, R. 2001. *Specification of Sustainability-Based Environmental Assessment Decision Criteria and Implications for Determining "Significance" in Environmental Assessment*. Ottawa: Canadian Environmental Assessment Agency. 42 p.

Gibson, R.B. 2006. Sustainability assessment: Basic components of a practical approach. *Impact Assessment and Project Appraisal* 24(3): 170–182.

Gilchrist, G., Heath, J., Arragutainaq, L., Robertson, G., Allard, K., Gilliland, S., and Mallory, M. 2006. Combining scientific and local knowledge to study common eider ducks wintering in Hudson Bay. In: *Climate Change: Linking Traditional and Scientific Knowledge.* Riewe, R. and Oakes, J., editors. Winnipeg: Aboriginal Issues Press. 189–201 p.

Gontier, M., Balfors, B., and Mörtberg, U. 2006. Biodiversity in environmental assessment – current practice and tools for prediction. *Environmental Impact Assessment Review* 26(3): 268–286.

Goodall, D.W. 1975. Ecosystem modeling in the desert biome. In: *Systems Analysis and Simulation in Ecology.* Volume III. Patten, B.C., editor. New York: Academic Press. 73–94 p.

Goodyear, C.P. 1985. Toxic materials, fishing, and environmental variation: Simulated effects on striped bass population trends. *Transactions of the American Fisheries Society* 114(1): 107–113.

Gosz, J.R., Likens, G.E., and Bormann, F.H. 1973. Nutrient release from decomposing leaf and branch litter in the Hubbard Brook Forest, New Hampshire. *Ecological Monographs* 43(2): 173–191.

Goward, S.N., and Hope, A.S. 1989. Evapotranspiration from combined reflected solar and emitted terrestrial radiation: preliminary FIFE results from AVHRR data. *Advances in Space Research* 9(7): 239–249.

Green, R.H. 1989. Power analysis and practical strategies for environmental monitoring. *Environmental Research* 50(1): 195–205.

Greig, L.A., and Duinker, P.N. 2011. A proposal for further strengthening science in environmental impact assessment in Canada. *Impact Assessment and Project Appraisal* 29(2): 159–165.

Greig, L.A., and Duinker, P.N. 2014. Strengthening impact assessment: What problems do integration and focus fix? *Impact Assessment and Project Appraisal* 32(1): 23–24.

Grimm, V. 1999. Ten years of individual-based modeling in ecology: What have we learned and what could we learn in the future? *Ecological Modeling* 115(2–3): 129–148.

Gustafson, E.J., Lytle, D.E., Swaty, R., and Loehle, C. 2007. Simulating the cumulative effects of multiple forest management strategies on landscape measures of forest sustainability. *Landscape Ecology* 22(1): 141–156.

Håkanson, L., and Carlsson, L. 1998. Fish farming in lakes and acceptable total phosphorus loads: Calibrations, simulations and predictions using the LEEDS model in Lake Southern Bullaren, Sweden. *Aquatic Ecosystem Health & Management* 1(1): 1–24.

Hall, R.J., Likens, G.E., Fiance, S.B., and Hendrey, G.R. 1980. Experimental acidification of a stream in the Hubbard Brook Experimental Forest, New Hampshire. *Ecology* 61(4): 976–989.

Handcock, R.N., Swain, D.L., Bishop-Hurley, G., Patison, K.P., Wark, T., Valencia, P., Corke, P., and O'Neill, C.J. 2009. Monitoring animal behavior and environmental interactions using wireless sensor networks, GPS collars and satellite remote sensing. *Sensors* 9(5): 3586–3603.

Hara, T.J., Law, Y.M.C., and Macdonald, S. 1976. Effects of mercury and copper on the olfactory response in rainbow trout, Salmo gairdneri. *Journal of Fisheries Research Board of Canada* 33(7): 1568–1573.

Hattula, M.L., Janatuinen, J., Särkkä, J., and Paasivirta, J. 1978. A five-year monitoring study of the chlorinated hydrocarbons in the fish of a Finnish lake ecosystem. *Environmental Pollution* 15(2): 121–139.

Haug, P., Burwell, R., Stein, A., and Bandurski, B. 1984. Determining the significance of environmental issues under the National Environmental Policy Act. *Journal of Environmental Management* 18: 15–24.

Heink, U., and Kowarik, I. 2010. What are indicators? On the definition of indicators in ecology and environmental planning. *Ecological Indicators* 10(3): 584–593.

Hendrey, G.R., Ellsworth, D.S., Lewin, K.F., and Nagy, J. 1999. A free-air enrichment system for exposing tall forest vegetation to elevated atmospheric CO_2. *Global Change Biology* 5(3): 293–309.

Herr, J.W., Chen, C.W., Goldstein, R.A., Herd, R., and Brown, J.M. 2003. Modeling acid mine drainage on a watershed scale for TMDL calculations. *Journal of the American Water Resources Association* 39(2): 289–300.

Herricks, E.E., Shanholtz, V.O., and Contractor, D.N. 1975. Models to predict environmental impact of mine drainage on streams. *Journal of American Soc Biol Eng* 18(4): 0657–0663.

Hewitt, L.M., Dubé, M.G., Ribey, S.C., Culp, J.M., Lowell, R., Hedley, K., Kilgour, B., Portt, C., MacLatchy, D.L., and Munkittrick, K.R. 2005. Investigation of cause in pulp and paper environmental effects monitoring. *Water Quality Research Journal of Canada* 40(3): 261–274.

Hewitt, L.M., Dubé, M.G., Culp, J.M., MacLatchy, D.L., and Munkittrick, K.R. 2003. A proposed framework for investigation of cause for environmental effects monitoring. *Human and Ecological Risk Assessment* 9(1): 195–211.

Hilborn, R., and Walters, C.J. 1981. Pitfalls of environmental baseline and process studies. *Environmental Impact Assessment Review* 2(3): 265–278.

Hilborn, R., and Walters, C.J., editors. 1992. *Quantitative Fisheries Stock Assessment: Choice, Dynamics and Uncertainty.* New York: Springer. 570 p.

Hill, C., Hill, S., Lamb, C., and Barrett, T.W. 1974. Sensitivity of native desert vegetation to SO_2 and to SO_2 and NO_2 Combined. *Journal of the Air Pollution Control Association* 24(2): 153–157.

Hill, J.G., and Price, D.R. 1983. The impact of deep mining on an overlying aquifer in western Pennsylvania. *Ground Water Monitoring & Remediation* 3(1): 138–143.

Hill, W.W., and Ortolano, L. 1978. NEPA's effect on the consideration of alternatives: A crucial test. *Natural Resources Journal* 18: 285–312.

Hindell, M.A. 1991. Some life-history parameters of a declining population of southern elephant seals, Mirounga leonina. *Journal of Animal Ecology* 60(1): 119–134.

Hirsch, A. 1980. The baseline study as a tool in environmental impact assessment. In: *Biological Evaluation of Environmental Impacts.* Proceedings of a Symposium at the 1976 Meeting of the Ecological Society of America. Washington, DC: Council on Environmental Quality. 84–93 p.

Hobbs, B.F. 1985. Choosing how to choose: Comparing amalgamation methods for environmental impact assessment. *Environmental Impact Assessment Review* 5(4): 301–319.

Holben, B.N. 1986. Characteristics of maximum-value composite images from temporal AVHRR data. *International Journal of Remote Sensing* 7(11): 1417–1434.

Hollick, M. 1981. The role of quantitative decision-making methods in environmental impact assessment. *Journal of Environmental Management* 12: 65–78.

Holling, C.S. 1973. Resilience and stability of ecological systems. *Annual Review of Ecology and Systematics* 4: 1–23.

Holling, C.S. 1986. The resilience of terrestrial ecosystems: Local surprise and global change. In: *Sustainable Development of the Biosphere*. Clark, W.C. and Munn, R.E., editors. Cambridge: Cambridge University Press. 292–317 p.

Holling, C.S., editor. 1978. *Adaptive Environmental Assessment and Management*. Toronto: John Wiley & Sons. 377 p.

Houghton, J.T., Jenkins, G.J., and Ephraums, J.J., editors. 1990. *Climate Change: The IPCC Scientific Assessment*. Cambridge: Cambridge University Press. 364 p.

Houle, M., Fortin, D., Dussault, C., Courtois, R., and Ouellet, J. 2010. Cumulative effects of forestry on habitat use by gray wolf (*Canis lupus*) in the boreal forest. *Landscape Ecology* 25(3): 419–433.

Hounslow, A.W. 1995. *Water Quality Data: Analysis and Interpretation*. Boca Raton: CRC Press. 397 p.

Hourcle, L.R. 1979. The new Council on Environmental Quality regulations: The tiger's new teeth. *A F L Rev* 21: 450–461.

Huete, A., Didan, K., Miura, T., Rodriguez, E.P., Gao, X., and Ferreira, L.G. 2002. Overview of the radiometric and biophysical performance of the MODIS vegetation indices. *Remote Sensing of Environment* 83(1–2): 195–213.

Huete, A.R. 1988. A soil-adjusted vegetation index (SAVI). *Remote Sensing of Environment* 25(3): 295–309.

Huntington, H., Callaghan, T., Fox, S., and Krupnik, I. 2004a. Matching traditional and scientific observations to detect environmental change: A discussion on arctic terrestrial ecosystems. *Ambio* 33(7): 18–23.

Huntington, H.P. 2000. Using traditional ecological knowledge in science: Methods and applications. *Ecological Applications* 10(5): 1270–1274.

Huntington, H.P., Suydam, R.S., and Rosenberg, D.H. 2004b. Traditional knowledge and satellite tracking as complementary approaches to ecological understanding. *Environmental Conservation* 31(3): 177–180.

Hushon, J.M. 1987. Expert systems for environmental problems. *Environmental Science & Technology* 21(9): 838–841.

Huston, M., DeAngelis, D., and Post, W. 1988. New computer models unify ecological theory. *Bioscience* 38(10): 682–691.

Huth, A., Drechsler, M., and Köhler, P. 2004. Multicriteria evaluation of simulated logging scenarios in a tropical rain forest. *Journal of Environmental Management* 71(4): 321–333.

Ijäs, A., Kuitunen, M.T., and Jalava, K. 2010. Developing the RIAM method (rapid impact assessment matrix) in the context of impact significance assessment. *Environmental Impact Assessment Review* 30(2): 82–89.

Innis, G.S. 1975. Role of total systems models in the grassland biome study. In: *Systems Analysis and Simulation in Ecology*. Volume III. Patten BC, editor. New York: Academic Press. 14–48 p.

Jackson, D.R., Selvidge, W.J., and Ausmus, B.S. 1978. Behavior of heavy metals in forest microcosms: II. effects on nutrient cycling processes. *Water, Air & Soil Pollution* 10(1): 13–18.

Jackson, D.R., Washburne, C.D., and Ausmus, B.S. 1977. Loss of Ca and NO_3-N from terrestrial microcosms as an indicator of soil pollution. *Water, Air & Soil Pollution* 8(3): 279–284.

Jager, H.I., Cardwell, H.E., Sale, M.J., Bevelhimer, M.S., Coutant, C.C., and Van Winkle, W. 1997. Modeling the linkages between flow management and salmon recruitment in rivers. *Ecological Modeling* 103(2): 171–191.

Janssen, P.H.M., and Heuberger, P.S.C. 1995. Calibration of process-oriented models. *Ecological Modeling* 83(1): 55–66.

Janssen, R. 2001. On the use of multi-criteria analysis in environmental impact assessment in the Netherlands. *Journal of Multi-Criteria Decision Analysis* 10(2): 101–109.

Jensen, A.L. 1982. Impact of a once-through cooling system on the yellow perch stock in the western basin of Lake Erie. *Ecological Modeling* 15(2): 127–144.

Jia, Y., Ni, G., Yoshitani, J., Kawahara, Y., and Kinouchi, T. 2002. Coupling simulation of water and energy budgets and analysis of urban development impact. *Journal of Hydrologic Engineering* 7(4): 302–311.

João, E. 2002. How scale affects environmental impact assessment. *Environmental Impact Assessment Review* 22(4): 289–310.

Johannes, R.E. 1993. Integrating traditional ecological knowledge and management with environmental impact assessment. In: *Traditional Ecological Knowledge: Concepts and Cases*. Inglis, J.T., editor. Ottawa: Canadian Museum of Nature. 33–39 p.

Johnson, C.J., Boyce, M.S., Case, R.L., Cluff, H.D., Gau, R.J., Gunn, A., and Mulders, R. 2005. Cumulative effects of human developments on Arctic wildlife. *Wildlife Monographs* 160: 1–36.

Johnson, G.D., Erickson, W.P., Strickland, M.D., Shepherd, M.F., Shepherd, D.A., Sarappo, S.A. 2003. Mortality of bats at a large-scale wind power development at Buffalo Ridge, Minnesota. *American Midland Naturalist* 150(2): 332–342.

Jones, C. 1999. Screening, scoping and the consideration of alternatives. In: *Handbook of Environmental Impact Assessment*. Volume 1. Petts, J., editor. Oxford: Blackwell Science. p. 201–228.

Jones, K.B., Bogena, H., Vereecken, H., and Weltzin, J.F. 2010. Design and importance of multi-tiered ecological monitoring networks. In: *Long-Term Ecological Research: Between Theory and Application*. Muller, F., Baessler, C., Schubert, H., et al, editors. Dordrecht: Springer. 355–374 p.

Jones, M., and Morrison-Saunders, A. 2016. Making sense of significance in environmental impact assessment. *Impact Assessment and Project Appraisal* 34(1): 87–93.

Jordan, D.N., Zitzer, S.F., Hendrey, G.R., Lewin, K.F., Nagy, J., Nowak, R.S., Smith, S.D., Coleman, J.S., and Seemann, J.R. 1999. Biotic, abiotic and performance aspects of the Nevada Desert Free-Air CO_2 Enrichment (FACE) facility. *Global Change Biology* 5(6): 659–668.

Jørgensen, S.E. 2008. Overview of the model types available for development of ecological models. *Ecological Modeling* 215(1–3): 3–9.

Jørgensen, S.E., and Bendoricchio, G. 2001. *Fundamentals of Ecological Modeling*. Third ed. Amsterdam: Elsevier. 628 p.

Justice, C.O., Giglio, L., Korontzi, S., Owens, J., Morisette, J.T., Roy, D., Descloitres, J., Alleaume, S., Petitcolin, F., and Kaufman, Y. 2002. The MODIS fire products. *Remote Sensing of Environment* 83(1–2): 244–262.

Justice, C.O., Vermote, E., Townshend, J.R.G., Defries, R., Roy, D.P., Hall, D.K., Salomonson, V.V., Privette, J.L., Riggs, G., Strahler, A., et al. 1998. The Moderate Resolution Imaging Spectroradiometer (MODIS): Land remote sensing for global change research. *IEEE Transactions on Geoscience and Remote Sensing* 36(4): 1228–1249.

Kain, J., and Söderberg, H. 2008. Management of complex knowledge in planning for sustainable development: The use of multi-criteria decision aids. *Environmental Impact Assessment Review* 28(1): 7–21.

Kalbar, P.P., Karmakar, S., and Asolekar, S.R. 2013. The influence of expert opinions on the selection of wastewater treatment alternatives: a group decision-making approach. *Journal of Environmental Management* 128: 844–851.

Kapos, V. 1989. Effects of isolation on the water status of forest patches in the Brazilian Amazon. *Journal of Tropical Ecology* 5(2): 173–185.

Karjalainen, T.P., Pussinen, A., Liski, J., Nabuurs, G., Erhard, M., Eggers, T., Sonntag, M., and Mohren, G.M.J. 2002. An approach towards an estimate of the impact of forest management and climate change on the European forest sector carbon budget: Germany as a case study. *Forest Ecology and Management* 162(1): 87–103.

Karkkainen, B.C. 2002. Toward a smarter NEPA: Monitoring and managing government's environmental performance. *Columbia Law Review* 102: 903–972.

Karp, J.P. 1978. Judicial review of environmental impact statement contents. *American Business Law Journal* 16(2): 127–156.

Karstens, S.A.M., Bots, P.W.G., and Slinger, J.H. 2007. Spatial boundary choice and the views of different actors. *Environmental Impact Assessment Review* 27(5): 386–407.

Kaspar, H.F., Hall, G.H., and Holland, A.J. 1988. Effects of sea cage salmon farming on sediment nitrification and dissimilatory nitrate reductions. *Aquaculture* 70(4): 333–344.

Kasper, R. 1977. Cost-benefit analysis in environmental decisionmaking. *The George Washington Law Review* 45(5): 1013–1024.

Kaufmann, R.F., Eadie, G.G., and Russell, C.R. 1976. Effects of uranium mining and milling on ground water in the Grants Mineral Belt, New Mexico. *Ground Water* 14(5): 296–308.

Kavvadias, V.A., Alifragis, D., Tsiontsis, A., Brofas, G., and Stamatelos, G. 2001. Litterfall, litter accumulation and litter decomposition rates in four forest ecosystems in northern Greece. *Forest Ecology and Management* 144(1–3): 113–127.

Kelly, M.G., Hornberger, G.M., and Cosby, B.J. 1974. Continuous automated measurement of rates of photosynthesis and respiration in an undisturbed river community. *Limnology and Oceanography* 19(2): 305–312.

Kelso, J.R.M. 1977. Density, distribution, and movement of Nipigon Bay fishes in relation to a pulp and paper mill effluent. *Journal of the Fisheries Research Board of Canada* 34(6): 879–885.

Kenchington, E.L.R., Gilkinson, K.D., MacIsaac, K.G., Bourbonnais-Boyce, C., Kenchington, T.J., Smith, S.J., Gordon, Jr D.C. 2006. Effects of experimental otter trawling on benthic assemblages on Western Bank, northwest Atlantic Ocean. *Journal of Sea Research* 56(3): 249–270.

Kennedy, A.J., and Ross, W.A. 1992. An approach to integrate impact scoping with environmental impact assessment. *Environmental Management* 16(4): 475–484.

Kenny, A.J., and Rees, H.L. 1996. The effects of marine gravel extraction on the macrobenthos: Results 2 years post-dredging. *Marine Pollution Bulletin* 32(8): 615–622.

Kessell, S.R., Good, R.B., and Hopkins, A.J.M. 1984. Implementation of two new resource management information systems in Australia. *Environmental Management* 8(3): 251–269.

Khadka, R.B., Mathema, A., and Shrestha, U.S. 2011. Determination of significance of environmental impacts of development projects: a case study of environmental impact assessment of Indrawati-3 Hydropower Project in Nepal. *Journal of Environmental Protection* 2(8): 1021–1031.

Kilgour, B.W., Dubé, M.G., Hedley, K., Portt, C.B., and Munkittrick, K.R. 2007. Aquatic environmental effects monitoring guidance for environmental assessment practitioners. *Environmental Monitoring and Assessment* 130(1): 423–436.

Kilgour, B.W., Dubé, M.G., Hedley, K., Portt, C.B., and Munkittrick, K.R. 2007. Aquatic environmental effects monitoring guidance for environmental assessment practitioners. *Environmental Monitoring and Assessment* 130(1): 423–436.

Kingston, R., Carver, S., Evans, A., and Turton, I. 2000. Web-based public participation geographical information systems: An aid to local environmental decision-making. *Computers Environment and Urban System* 24(2): 109–125.

Kjellerup, U. 1999. Significance determination: A rational reconstruction of decisions. *Environmental Impact Assessment Review* 19(1): 3–19.

Knapp, A.K., and Seastedt, T.R. 1986. Detritus accumulation limits productivity of tall grass prairie. *Bioscience* 36(10): 662–668.

Konisky, D.M., and Bierle, T.C. 2001. Innovations in public participation and environmental decision making: Examples from the Great Lakes region. *Society & Natural Resources* 14(9): 815–826.

Koop, K., Booth, D., Broadbent, A., and Brodie, J., Bucher, D., Capone, D., Coll, J., Dennison, W., Erdmann, M., Harrison, P., et al. 2001. ENCORE: the effect of nutrient enrichment on coral reefs. Synthesis of Results and Conclusions. *Marine Pollution Bulletin* 42(2): 91–120.

Koterba, M.T., Hornbeck, J.W., and Pierce, R.S. 1979. Effects of sludge applications on soil water solution and vegetation in a northern hardwood stand. *Journal of Environmental Quality* 8(1): 72–78.

Krause, H.H. 1982. Nitrate formation and movement before and after clear-cutting of a monitored watershed in central New Brunswick, Canada. *Canadian Journal of Forest Research* 12(4): 922–930.

Krausman, P.R., Leopold, B.D., and Scarbrough, D.L. 1986. Desert mule deer response to aircraft. *Wildlife Society B* 14(1): 68–70.

Kremling, K., Piuze, J., von Bröckel, K., and Wong, C.S. 1978. Studies on the pathways and effects of cadmium in controlled ecosystem encolosures. *Marine Biology* 48(1): 1–10.

Kurz, W.A., and Apps, M.J. 2006. Developing Canada's national forest carbon monitoring, accounting and reporting system to meet the reporting requirements of the Kyoto Protocol. *Mitigation and Adaptation Strategies for Global Change* 11(1): 33–43.

Kurz, W.A., Dymond, C.C., Stinson, G., Rampley, G.J., Neilson, E.T., Carroll, A.L., Ebata, T., and Safranyik, L. 2008. Mountain pine beetle and forest carbon feedback to climate change. *Nature* 452(7190): 987–990.

Kurz, W.A., Dymond, C.C., White, T.M., Stinson, G., Shaw, C.H., Rampley, G.J., Smyth, C., Simpson, B.N., Neilson, E.T., Trofymow, J.A., et al. 2009. CBM-CFS3: A model of carbon-dynamics in forestry and land-use change implementing IPCC standards. *Ecological Modeling* 220(4): 480–504.

La Rosa, T., Mirto, S., Favaloro, E., Savona, B., Sarà, G., Danovaro, R., and Mazzola, A. 2002. Impact on the water column biogeochemistry of a Mediterranean mussel and fish farm. *Water Research* 36(3): 713–721.

Langdon, C., Broecker, W.S., Hammond, D.E., Glenn, E., Fitzsimmons, K., Nelson, S.G., Peng, T., Hajdas, I., and Bonani, G. 2003. Effect of elevated CO_2 on the community metabolism of an experimental coral reef. *Global Biogeochemical Cycles* 17(1): 11–1, 11–14.

Lapping, M.B. 1975. Environmental impact assessment methodologies: A critique. *Boston College Environmental Affairs Law Review* 4(1): 123–134.

Larsen, S.V., Kørnøv, L., and Driscoll, P. 2013. Avoiding climate change uncertainties in strategic environmental assessment. *Environmental Impact Assessment Review* 43: 144–150.

Larson, M.A., Thompson III, F.R., Millspaugh, J.J., Dijak, W.D., and Shifley, S.R. 2004. Linking population viability, habitat suitability, and landscape simulation models for conservation planning. *Ecological Modeling* 180(1): 103–118.

Lawe, L.B., Wells, J., and Cree, M. 2005. Cumulative effects assessment and EIA follow-up: A proposed community-based monitoring program in the Oil Sands Region, Northeastern Alberta. *Impact Assessment Project Appraisal* 23(3): 205–209.

Lawrence, D.P. 1993. Quantitative versus qualitative evaluation: A false dichotomy? *Environmental Impact Assessment Review* 13(1): 3–11.

Lawrence, D.P. 2007a. Impact significance determination – designing an approach. *Environmental Impact Assessment Review* 27(8): 730–754.

Lawrence, D.P. 2007b. Impact significance determination – back to basics. *Environmental Impact Assessment Review* 27(8): 755–769.

Lawrence, D.P. 2007c. Impact significance determination – pushing the boundaries. *Environmental Impact Assessment Review* 27(8): 770–788.

Leach, J.M., and Thakore, A.N. 1975. Isolation and identification of constituents toxic to juvenile rainbow trout (*Salmo gairdneri*) in caustic extraction effluents from kraft pulpmill bleach plants. *Journal of Fisheries Research Board of Canada* 32(8): 1249–1257.

Lee, K.N. 1993. *Compass and Gyroscope: Integrating Science and Politics for the Environment*. Washington, DC: Island Press. 243 p.

Lee, N., and Brown, D. 1992. Quality control in environmental assessment. *Project Appraisal* 7(1): 41–45.

Lee, N., and Colley, R. 1991. Reviewing the quality of environmental statements: Review methods and findings. *Town Planning Review* 62(2): 239–248.

Lees, J., Jaeger, J.A.G., Gunn, J.A.E., and Noble, B.F. 2016. Analysis of uncertainty consideration in environmental assessment: an empirical study of Canadian

EA practice. *Journal of Environmental Planning and Management* 59(11): 2024–2044.

Lefsky, M.A., Cohen, W.B., Parker, G.G., and Harding, D.J. 2002. Lidar remote sensing for ecosystem studies. *Bioscience* 52(1): 19–30.

Lehodey, P., Senina, I., Calmettes, B., Hampton, J., and Nicol, S. 2013. Modeling the impact of climate change on Pacific skipjack tuna population and fisheries. *Climatic Change* 119(1): 95–109.

Lein, J.K. 1989. An expert system approach to environmental impact assessment. *International Journal of Environmental Studies* 33(1–2): 13–27.

Leknes, E. 2001. The roles of EIA in the decision-making process. *Environmental Impact Assessment Review* 21(4): 309–334.

Leopold, L.B., Clarke, F.E., Hanshaw, B.B., and Balsley, J.R. 1971. *A Procedure for Evaluating Environmental Impact.* Washington, DC: US Department of the Interior. 13 p.

Leslie, J.K., and Kelso, J.R.M. 1977. Influence of a pulp and paper mill effluent on aspects of distribution, survival and feeding of Nipigon Bay, Lake Superior, larval fish. *Bulletin of Environmental Contamination and Toxicology* 18(5): 602–610.

Lett, P.F., Beamish, F.W.H., and Farmer, G.J. 1975. System simulation of the predatory activities of sea lampreys (Petromyzon marinus) on lake trout (*Salvelinus namaycush*). *Journal of Fisheries Research Board of Canada* 32(5): 623–631.

Leung, W., Noble, B.F., Jaeger, J.A.G., and Gunn, J.A.E. 2016. Disparate perceptions about uncertainty consideration and disclosure practices in environmental assessment and opportunities for improvement. *Environmental Impact Assessment Review* 57: 89–100.

Leung, W., Noble, B., Gunn, J., and Jaeger, J.A.G. 2015. A review of uncertainty research in impact assessment. *Environmental Impact Assessment Review* 50: 116–123.

Leventhal, H. 1974. Environmental decision making and the role of the courts. *University of Pennsylvania Law Review* 122(3): 509–555.

Likens, G.E. 1983. Address of the past president: Grand Forks, North Dakota; August 1983: A priority for ecological research. *Bulletin of Ecological Society of America* 64(4): 234–243.

Likens, G.E., Bormann, F.H., Johnson, N.M., Fisher, D.W., and Pierce, R.S. 1970. Effects of forest cutting and herbicide treatment on nutrient budgets in the Hubbard Brook watershed-ecosystem. *Ecological Monographs* 40(1): 23–47.

Lindegren, M., Möllmann, C., Nielsen, A., Brander, K., MacKenzie, B.R., and Stenseth, N.C. 2010. Ecological forecasting under climate change: the case of Baltic cod. *Proceedings Biological Sciences* 277(1691): 2121–2130.

Lowell, R.B., Culp, J.M., and Dubé, M.G. 2000. A weight-of-evidence approach for Northern River risk assessment: Integrating the effects of multiple stressors. *Environmental Toxicology and Chemistry* 19(4): 1182–1190.

Lynch, R.M. 1972. Complying with NEPA: The tortuous path to an adequate environmental impact statement. *Arizona Law Review* 14(4): 717–745.

MacArthur, R.A., Johnston, R.H., and Geist, V. 1979. Factors influencing heart rate in free-ranging bighorn sheep: A physiological approach to the study of wildlife harassment. *Canadian Journal of Zoology* 57(10): 2010–2021.

Magnuson, J.J., Benson, B.J., and Kratz, T.K. 1990. Temporal coherence in the limnology of a suite of lakes in Wisconsin, U.S.A. *Freshwater Biology* 23(1): 145–159.

Malcolm, J.R. 1988. Small mammal abundances in isolated and non-isolated primary forest reserves near Manaus, Brazil. *Acta Amazonica* 18: 67–83.

Malingreau, J.P., Tucker, C.J., and Laporte, N. 1989. AVHRR for monitoring global tropical deforestation. *International Journal of Remote Sensing* 10(4–5): 855–867.

Mango, L.M., Melesse, A.M., McClain, M.E., Gann, D., and Setegn, S.G. 2011. Land use and climate change impacts on the hydrology of the upper Mara River basin, Kenya: Results of a modeling study to support better resource management. *Hydrology and Earth System Sciences* 15(7): 2245–2258.

Manly, B.F.J., McDonald, L.L., Thomas, D.L., McDonald, T.L., and Erickson, W.P. 2002. *Resource Selection by Animals: Statistical Design and Analysis for Field Studies.* Second ed. New York: Kluwer Academic Publishers. 221 p.

Mardle, S., Pascoe, S., and Herrero, I. 2004. Management objective importance in fisheries: An evaluation using the analytic hierarchy process (AHP). *Environmental Management* 33(1): 1–11.

Marshall, R. 2005. Environmental impact assessment follow-up and its benefits for industry. *Impact Assessment and Project Appraisal* 23(3): 191–196.

Marshall, R., Arts, J., and Morrison-Saunders, A. 2005. International principles for best practice EIA follow-up. *Impact Assessment Project Appraisal* 23(3): 175–181.

Marzluff, J.M., Millspaugh, J.J., Ceder, K.R., Oliver, C.D., Withey, J., McCarter, J.B., Mason, C.L., and Comnick, J. 2002. Modeling changes in wildlife habitat and timber revenues in response to forest management. *Journal of Forest Science* 48(2): 191–202.

Mathur, D., Heisey, P.G., and Magnusson, N.C. 1977. Impingement of fishes at peach bottom atomic power station, Pennsylvania. *Transactions of the American Fisheries Society* 106(3): 258–267.

Mattson, D.J., Knight, R.R., and Blanchard, B.M. 1987a. The effects of developments and primary roads on grizzly bear habitat use in Yellowstone National Park, Wyoming. *Bears: Their Biology and Management* 7: 259–273.

Mattson, K.G., Swank, W.T., and Waide, J.B. 1987b. Decomposition of woody debris in a regenerating, clear-cut forest in the southern Appalachians. *Canadian Journal of Forest Research* 17(7): 712–721.

May, R.M. 1973. *Stability and Complexity in Model Ecosystems.* Princeton: Princeton University Press. 235 p.

Maynard-Smith, J. 1974. *Models in Ecology.* Cambridge: Cambridge University Press. 145 p.

McAllister, D.M. 1980. *Evaluation in Environmental Planning.* Cambridge, MA: MIT Press. 308 p.

McBroom, M., Thomas, T., and Zhang, Y. 2012. Soil erosion and surface water quality impacts of natural gas development in East Texas, USA. *Water* 4(4): 944–958.

McCaffery, K.R. 1973. Road-kills show trends in Wisconsin deer populations. *Journal of Wildlife Management* 37(2): 212–216.

McCallum, H. 2000. *Population Parameters: Estimation for Ecological Models.* Oxford: Blackwell Science. 348 p.

McClanahan, T.R. 1995. A coral reef ecosystem-fisheries model: Impacts of fishing intensity and catch selection on reef structure and processes. *Ecological Modeling* 80(1): 1–19.

McCleave, J.D. 2001. Simulation of the impact of dams and fishing weirs on reproductive potential of silver-phase American eels in the Kennebec River basin, Maine. *North American Journal of Fisheries Management* 21(3): 592–605.

McDonald, G.T., and Brown, L. 1995. Going beyond environmental impact assessment: Environmental input to planning and design. *Environmental Impact Assessment Review* 15(6): 483–495.

McKay, L.D., Cherry, J.A., and Gillham, R.W. 1993. Field experiments in a fractured clay till: 1. hydraulic conductivity and fracture aperture. *Water Resources Research* 29(4): 1149–1162.

McKelvey, K.B., Noon, B.R., and Lamberson, R.H. 1992. Conservation planning for species occupying fragmented landscapes: The case of the northern spotted owl. In: *Biotic Interactions and Global Change.* Kareiva, P.M., Kingsolver, J.G., and Huey, R.B., editors. Boston: Sinauer. 424–450 p.

McLellan, B.N., and Shackleton, D.M. 1988. Grizzly bears and resource-extraction industries: Effects of roads on behavior, habitat use and demography. *Journal of Applied Ecology* 25(2): 451–460.

McRae, B.H., Schumaker, N.H., McKane, R.B., Busing, R.T., Solomon, A.M., and Burdick, C.A. 2008. A multi-model framework for simulating wildlife population response to land-use and climate change. *Ecological Modeling* 219(1–2): 77–91.

Menzel, D.W., and Case, J. 1977. Concept and design: Controlled ecosystem pollution experiment. *Bulletin of Marine Science* 27(1): 1–7.

Merriman, D., and Thorpe, L.M., editors. 1976. *The Conneticut River Ecological Study: The Impact of a Nuclear Power Plant.* Washington, DC: American Fisheries Society. 252 p.

Miller, A., and Cuff, W. 1986. The Delphi approach to the mediation of environmental disputes. *Environmental Management* 10(3): 321–330.

Miller, P.C., Collier, B.D., and Bunnell, F.L. 1975. Development of ecosystem modeling in the tundra biome. In: *Systems Analysis and Simulation in Ecology.* Volume III. Patten, B.C., editor. New York: Academic Press. 95–116 p.

Mooij, W.M., and DeAngelis, D.L. 1999. Error propagation in spatially explicit population models: A reassessment. *Conservation Biology* 13(4): 930–933.

Moore, N.W. 1966. A pesticide monitoring system with special reference to the selection of indicator species. *Journal of Applied Ecology* 3: 261–269.

Morrison, M.L., Marcot, B.G., and Mannan, R.W. 2006. *Wildlife-Habitat Relationships: Concepts and Applications.* Third ed. Washington: Island Press. 493 p.

Morrison-Saunders, A., and Arts, J. 2004. *Assessing Impact: Handbook of EIA and SEA Follow-Up.* London: Earthscan. 338 p.

Morrison-Saunders, A, Arts J. 2005. Learning from experience: Emerging trends in environmental impact assessment follow-up. *Impact Assessment Project Appraisal* 23(3): 170–174.

122   *Science in the EIA process*

Morrison-Saunders, A., and Bailey, J. 2000. Transparency in environment impact assessment decision-making: Recent developments in Western Australia. *Impact Assessment Project Appraisal* 18(4): 260–270.

Morrison-Saunders, A., Pope, J., Gunn, J.A.E., Bond, A., and Retief, F. 2014. Strengthening impact assessment: A call for integration and focus. *Impact Assessment Project Appraisal* 32(1): 2–8.

Mulvihill, P.R. 2003. Expanding the scoping community. *Environmental Impact Assessment Review* 23(1): 39–49.

Mulvihill, P.R., and Baker, D.C. 2001. Ambitious and restrictive scoping: Case studies from Northern Canada. *Environmental Impact Assessment Review* 21(4): 363–384.

Mulvihill, P.R., and Jacobs, P. 1998. Using scoping as a design process. *Environmental Impact Assessment Review* 18(4): 351–369.

Munda, G. 1996. Cost-benefit analysis in integrated environmental assessment: Some methodological issues. *Ecological Economics* 19(2): 157–168.

Munkittrick, K.R., McGeachy, A., McMaster, M.E., and Courtenay, S.C. 2002. Overview of freshwater fish studies from the pulp and paper environmental effects monitoring program. *Water Quality Research Journal of Canada* 37(1): 49–77.

Munn, R.E., editor. 1979. *Environmental Impact Assessment: Principles and Procedures*. Second ed. Toronto: John Wiley & Sons. 190 p.

Munson, T.O., and Huggett, R.J. 1972. Current status of research on the biological effects of pesticides in Chesapeake Bay. *Chesapeake Science* 13(1): S154–S156.

Murdoch, L.C. 1992. Hydraulic fracturing of soil during laboratory experiments Part 1. Methods and observations. *Géotechnique* 43(2): 255–265.

Murkin, H.R., and Kadlec, J.A. 1986. Relationships between waterfowl and macro invertebrate densities in a northern prairie marsh. *Journal of Wildlife Management* 50(2): 212–217.

Murkin, H.R., Murkin, E.J., and Ball, J.P. 1997. Avian habitat selection and prairie wetland dynamics: A 10-year experiment. *Ecological Applications* 7(4): 1144–1159.

NAS (National Academy of Sciences). 1975. *Productivity of World Ecosystems*. Washington, DC: National Academy of Sciences. 166 p.

NASA (National Aeronautics and Space Administration). 1984. *Technical Memorandum 86129. Earth Observing System. NASA-TM-86129-VOL-1-PT-1*. Greenbelt: National Aeronautics and Space Administration. 51 p.

Nayak, S., Pamdeya, A., Gupta, M.C., Trivedi, C.R., Prasad, K.N., and Kadri, S.A. 1989. Application of satellite data for monitoring degradation of tidal wetlands of the Gulf of Kachchh, Western India. *Acta Astronaut* 20: 171–178.

Nelson, J.G., and Serafin, R. 1991. *Biodiversity and Environmental Assessment*. Waterloo: International Union for the Conservation of Nature. 13 p.

Nicholls, T.H., and Warner, D.W. 1972. Barred owl habitat use as determined by radiotelemetry. *Journal of Wildlife Management* 36(2): 213–224.

Niehoff, D., Fritsch, U., and Bronstert, A. 2002. Land-use impacts on storm-runoff generation: Scenarios of land-use change and simulation of hydrological response in a meso-scale catchment in SW-Germany. *Journal of Hydrology* 267(1–2): 80–93.

Nijkamp, P.1986. Multiple criteria analysis and integrated impact analyis. *Impact Assessment* 4(3–4): 226–261.

Nikolaidis, N.P., Rajaram, H., Schnoor, J.L., and Georgakakos, K.P. 1988. A generalized soft water acidification model. *Water Resources Research* 24(12): 1983–1996.

Noble, B., and Birk, J. 2011. Comfort monitoring? Environmental assessment follow-up under community – industry negotiated environmental agreements. *Environmental Impact Assessment Review* 31(1): 17–24.

Noble, B.F. 2000. Strengthening EIA through adaptive management: A systems perspective. *Environmental Impact Assessment Review* 20(1): 97–111.

Noble, B., and Storey, K. 2005. Towards increasing the utility of follow-up in Canadian EIA. *Environmental Impact Assessment Review* 25(2): 163–180.

NRC (National Research Council). 1986a. *Ecological Knowledge and Environmental Problem-Solving: Concepts and Case Studies.* Washington, DC: National Academy Press. 388 p.

NRC (National Research Council). 1986b. *Global Change in the Geosphere-Biosphere: Initial Priorities for an IGBP.* Washington, DC: National Academy Press. 91 p.

O'Faircheallaigh, C. 2007. Environmental agreements, EIA follow-up and aboriginal participation in environmental management: the Canadian experience. *Environmental Impact Assessment Review* 27(4): 319–342.

Olmstead, S.M., Muehlenbachs, L.A., Shih, J., Chu, Z., and Krupnick, A.J. 2013. Shale gas development impacts on surface water quality in Pennsylvania. *Proceedings of the National Academy of Sciences of the USA* 110(13): 4962–4967.

Olness, A., Smith, S.J., Rhoades, E.D., and Menzel, R.G. 1975. Nutrient and sediment discharge from agricultural watersheds in Oklahoma. *Journal of Environmental Quality* 4(3): 331–336.

O'Neill, R.V. 1975. Modeling in the eastern deciduous forest biome. In: *Systems Analysis and Simulation in Ecology.* Volume III. Patten, B.C., editor. New York: Academic Press. 49–72 p.

Onstad, C.A., and Foster, G.R. 1975. Erosion modeling on a watershed. *Transaction of American Society for Biological Engineering* 18(2): 288–292.

Ortolano, L., and Shepherd, A. 1995. Environmental impact assessment: Challenges and opportunities. *Impact Assess* 13(1): 3–30.

Ottar, B. 1976. Organization of long range transport of air pollution monitoring in Europe. *Water, Air, & Soil Pollution* 6(2): 219–229.

Overton, W.S. 1975. The ecosystem modeling approach in the coniferous forest biome. In: *Systems Analysis and Simulation in Ecology.* Volume III. Patten, B.C., editor. New York: Academic Press. 117–138 p.

Parisé, J., and Walker, T.R. 2017. Industrial wind turbine post-construction bird and bat monitoring: A policy framework for Canada. *Journal of Environmental Management* 201: 252–259.

Park, R.A. 1974. A generalized model for simulating lake ecosystems. *Simulation* 23(2): 33–50.

Pastor, J., and Post, W.M. 1988. Response of northern forests to CO_2-induced climate change. *Nature* 334(6177): 55–58.

124   *Science in the EIA process*

Patten, B.C. 1971. A primer for ecological modeling and simulation with analog and digital computers. In: *Systems Analysis and Simulation in Ecology.* Volume I. Patten, B.C., editor. New York: Academic Press. 3–121 p.

Pearce, D. 1976. The limits of cost-benefit analysis as a guide to environmental policy. *Kyklos* 29(1): 97–112.

Pearlstine, L., McKellar, H., and Kitchens, W. 1985. Modeling the impacts of a river diversion on bottomland forest communities in the Santee River floodplain, South Carolina. *Ecological Modeling* 29(1): 283–302.

Pellew, R.A.P. 1983. The impacts of elephant, giraffe and fire upon the Acacia tortilis woodlands of the Serengeti. *African Journal of Ecology* 21(1): 41–74.

Peterman, R.M. 1975. New techniques for policy evaluation in ecological systems: Methodology for a case study of Pacific salmon fisheries. *Journal of Fisheries Research Board of Canada* 32(11): 2179–2188.

Peterson, E.B., Chan, Y.H., Peterson, N.M., Constable, G.A., Caton, R.B., Davis, C.S., Wallace, R.R., and Yarranton, G.A. 1987. *Cumulative Effects Assessment in Canada: An Agenda for Action and Research.* Hull: Canadian Environmental Assessment Research Council. 63 p.

Petts, J. 1999. Public participation and environmental impact assessment. In: *Handbook of Environmental Impact Assessment.* Volume 1. Petts, J., editor. Oxford: Blackwell Science. p. 145–177.

Petts, J. 2003. Barriers to deliberative participation in EIA: Learning from waste policies, plans and projects. *Journal of Environmental Assessment Policy and Management* 5(3): 269–293.

Pielou, E.C. 1977. *Mathematical Ecology.* New York: John Wiley & Sons. 385 p.

Pischke, F., and Cashmore, M. 2006. Decision-oriented environmental assessment: An empirical study of its theory and methods. *Environmental Impact Assessment Review* 26(7): 643–662.

Plonczkier, P., and Simms, I.C. 2012. Radar monitoring of migrating pink-footed geese: Behavioral responses to offshore wind farm development. *Journal of Applied Ecology* 49(5): 1187–1194.

Põder, T., and Lukki, T. 2011. A critical review of checklist-based evaluation of environmental impact statements. *Impact Assessment Project Appraisal* 29(1): 27–36.

Polfus, J.L., Hebblewhite, M., and Heinemeyer, K. 2011. Identifying indirect habitat loss and avoidance of human infrastructure by northern mountain woodland caribou. *Biological Conservation* 144(11): 2637–2646.

Polgar, T.T., Turner, M.A., and Summers, J.K. 1988. Effect of power plant entrainment on the population dynamics of the bay anchovy (*Anchoa mitchilli*). *Ecological Modeling* 41(3): 201–218.

Prescott, C.E., Blevins, L.L., and Staley, C.L. 2000. Effects of clear-cutting on decomposition rates of litter and forest floor in forests of British Columbia. *Canadian Journal of Forest Research* 30(11): 1751–1757.

Priede, I.G. 1984. A basking shark (*Cetorhinus maximus*) tracked by satellite together with simultaneous remote sensing. *Fisheries Research* 2(3): 201–216.

Priede, I.G., and Young, A.H. 1977. The ultrasonic telemetry of cardiac rhythms of wild brown trout (Salmo trutta L.) as an indicator of bio-energetics and behavior. *Journal of Fish Biology* 10(4): 299–318.

Pulliam, H.R., Dunning, J.B., and Liu, J. 1992. Population dynamics in complex landscapes: A case study. *Ecological Applications* 2(2): 165–177.

Quick, M.C., and Pipes, A. 1976. A combined snowmelt and rainfall runoff model. *Canadian Journal of Civil Engineering* 3(3): 449–460.

Railsback, S.F., and Harvey, B.C. 2002. Analysis of habitat-selection rules using an individual-based model. *Ecology* 83(7): 1817–1830.

Ramanathan, R. 2001. A note on the use of the analytic hierarchy process for environmental impact assessment. *Journal of Environmental Management* 63(1): 27–35.

Ray, S., and White, W. 1976. Selected aquatic plants as indicator species for heavy metal pollution. *Journal of Environmental Science and Health A* 11(12): 717–725.

Reagan, M.T., Moridis, G.J., Keen, N.D., and Johnson, J.N. 2015. Numerical simulation of the environmental impact of hydraulic fracturing of tight/shale gas reservoirs on near-surface groundwater: background, base cases, shallow reservoirs, short-term gas, and water transport. *Water Resources Research* 51(4): 2543–2573.

Reed, M., French, D.P., Calambokidis, J., and Cubbage, J.C. 1989. Simulation modeling of the effects of oil spills on population dynamics of northern fur seals. *Ecological Modeling* 49(1): 49–71.

Reed, M., Spaulding, M.L., Lorda, E., Walker, H., and Saila, S.B. 1984. Oil spill fishery impact assessment modeling: The fisheries recruitment problem. *Estuarine, Coastal and Shelf Science* 19(6): 591–610.

Refsgaard, J.C., van der Sluijs, J.P., Højberg, A.L., and Vanrolleghem, P.A. 2007. Uncertainty in the environmental modeling process – a framework and guidance. *Environmental Modeling Software* 22(11): 1543–1556.

Reich, P.B., Knops, J., Tilman, D., Craine, J., Ellsworth, D., Tjoelker, M., Lee, T., Wedin, D., Naeem, S., Bahauddin, D., et al. 2001. Plant diversity enhances ecosystem responses to elevated CO_2 and nitrogen deposition. *Nature* 410(6830): 809–810.

Reimers, W. 1990. Estimating hydrological parameters from basin characteristics for large semiarid catchments. In: *Regionalization in Hydrology. Proceedings of the Ljubljana Symposium*. Wallingford: International Association of Hydrological Sciences. 187 p.

Ribey, S.C., Munkittrick, K.R., McMaster, M.E., Courtenay, S., Langlois, C., Munger, S., Rosaasen, A., and Whitley, G. 2002. Development of a monitoring design for examining effects in wild fish associated with discharges from metal mines. *Water Quality Research Journal of Canada* 37(1): 229–249.

Richardson, T. 2005. Environmental assessment and planning theory: four short stories about power, multiple rationality, and ethics. *Environmental Impact Assessment Review* 25(4): 341–365.

Richey, J.S., Mar, B.W., and Horner, R.R. 1985. Delphi technique in environmental assessment. I. implementation and effectiveness. *Journal of Environmental Economics and Management* 21: 135–146.

Richter, O., and Sondgerath, D. 1990. *Parameter Estimation in Ecology: The Link Between Data and Models*. New York: VCH Publishers. 218 p.

Roach, B., and Walker, T.R. 2017. Aquatic monitoring programs conducted during environmental impact assessments in Canada: Preliminary assessment before and after weakened environmental regulation. *Environmental Monitoring and Assessment* 189: 109.

Rock, B.N., Vogelmann, J.E., Williams, D.L., Vogelmann, A.F., and Hoshizaki, T. 1986. Remote detection of forest damage. *Bioscience* 36(7): 439–445.

Rodgers, A.R. 2001. Recent telemetry technology. In: *Radio Tracking and Animal Populations*. Millspaugh, J.J. and Marzluff, J.M., editors. San Diego: Academic Press. 79–121 p.

Rooney, R.C., and Podemski, C.L. 2009. Effects of an experimental rainbow trout (*Oncorhynchus mykiss*) farm on invertebrate community composition. *Canadian Journal of Fisheries and Aquatic Sciences* 66(11): 1949–1964.

Rosenberg, D.M., Resh, V.H., Balling, S.S., Barnby, M.A., Collins, J.N., Durbin, D.V., Flynn, T.S., Hart, D.D., Lamberti, G.A., McElravy, E.P., et al. 1981. Recent trends in environmental impact assessment. *Canadian Journal of Fisheries and Aquatic Sciences* 38(5): 591–624.

Rosenzweig, C., Solecki, W.D., Cox, J., Hodges, S., Parshall, L., Lynn, B., Goldberg, R., Gaffin, S., Slosberg, R.B., Savio, P., et al. 2009. Mitigating New York City's heat island: integrating stakeholder perspectives and scientific evaluation. *Bulletin of the American Meteorological Society* 90(9): 1297–1312.

Ross, W.A. 1987. Evaluating environmental impact statements. *Journal of Environmental Management* 25: 137–147.

Ross, W.A. 2000. Reflections of an environmental assessment panel member. *Environmental Impact Assessment Review* 18(2): 91–98.

Ross, W.A., Morrison-Saunders, A., Marshall, R., Sánchez, L.E., Weston, J., Au, E., Morgan, R.K., Fuggle, R., Sadler, B., Ross, W.A., et al. 2006. Improving quality. *Impact Assessment Project Appraisal* 24(1): 3–22.

Rossini, F.A., and Porter, A.L. 1982. Public participation and professionalism in impact assessment. *Nonprofit and Voluntary Sector Quarterly Sec Q* 11(1): 24–33.

Rouse, J., Haas, R., Schell, J., Deering, D., and Harlan, J. 1973. *Monitoring the Vernal Advancements and Retrogradation* (Greenwave Effect) of Nature Vegetation. NASA/GSFC Final Report. Greenbelt: National Aeronautics and Space Administration. 93 p.

Roux, D.J., Kempster, P.L., Kleynhans, C.J., Van Vliet, H.R., and Du Preez, H.H. 1999. Integrating stressor and response monitoring into a resource-based water-quality assessment framework. *Environmental Management* 23(1): 15–30.

Ruckelshaus, M., Hartway, C., and Kareiva, P. 1997. Assessing the data requirements of spatially explicit dispersal models. *Conservation Biology* 11(6): 1298–1306.

Rulifson, R.A. 1977. Temperature and water velocity effects on the swimming performances of young-of-the-year striped mullet (*Mugil cephalus*), spot (*Leiostomus xanthurus*), and pinfish (*Lagodon rhomboides*). *Journal of Fisheries Research Board of Canada* 34(12): 2316–2322.

Rundel, P.W., Graham, E.A., Allen, M.F., Fisher, J.C., and Harmon, T.C. 2009. Environmental sensor networks in ecological research. *New Phytologist* 182(3): 589–607.

Running, S.W., and Coughlan, J.C. 1988. A general model of forest ecosystem processes for regional applications I: Hydrologic balance, canopy gas exchange and primary production processes. *Ecological Modeling* 42(2): 125–154.

Rylands, A.B., and Keuroghlian, A. 1988. Primate populations in continuous forest and forest fragments in central Amazonia. *Acta Amazonica* 18: 291–307.

Sadler, B. 1996. *Environmental Assessment in a Changing World: Evaluating Practice to Improve Performance.* Final Report of the International Study of the Effectiveness of Environmental Assessment. Hull: Canadian Environmental Assessment Agency. 248 p.

Sallenave, J. 1994. Giving traditional ecological knowledge its rightful place in environmental impact assessment. *Northern Perspectives* 22(1): 16–19.

Sandström, O., and Thoresson, G. 1988. Mortality in perch populations in a Baltic pulp mill effluent area. *Marine Pollution Bulletin* 19(11): 564–567.

Sassaman, R.W. 1981. Threshold of concern: A technique for evaluation environmental impacts and amenity values. *J For.* 79(2): 84–86.

Scheller, R.M., Domingo, J.B., Sturtevant, B.R., Williams, J.S., Rudy, A., Gustafson, E.J., and Mladenoff, D.J. 2007. Design, development, and application of LANDIS-II, a spatial landscape simulation model with flexible temporal and spatial resolution. *Ecological Modeling* 201(3–4): 409–419.

Scheller, R.M., and Mladenoff, D.J. 2004. A forest growth and biomass module for a landscape simulation model, LANDIS: design, validation, and application. *Ecological Modeling* 180(1): 211–229.

Scheller, R.M., and Mladenoff, D.J. 2005. A spatially interactive simulation of climate change, harvesting, wind, and tree species migration and projected changes to forest composition and biomass in northern Wisconsin, USA. *Global Change Biology* 11(2): 307–321.

Schiffer, R., and Rossow, W.B. 1983. The International Satellite Cloud Climatology Project (ISCCP): The first project of the World Climate Research Programme. *Bulletin of the American Meteorological Society* 64(7): 779–784.

Schindler, D.W. 1976. The impact statement boondoggle. *Science* 192(4239): 509.

Schindler, D.W., Armstrong, F.A.J., Holmgren, S.K., and Brunskill, G.J. 1971. Eutrophication of Lake 227, Experimental Lakes Area, Northwestern Ontario, by addition of phosphate and nitrate. *Journal of Fisheries Research Board of Canada* 28(11): 1763–1782.

Schindler, D.W., Wagemann, R., Cook, R.B., Ruszczynski, T., and Prokopowich, J. 1980. Experimental acidification of Lake 223, Experimental Lakes Area: background data and the first three years of acidification. *Canadian Journal of Fisheries and Aquatic Sciences* 37(3): 342–354.

Schlesinger, W.H. 1977. Carbon balance in terrestrial detritus. *Annual Review of Ecology and Systematics* 8: 51–81.

Schmidt, M., Glasson, J., Emmelin, L., and Helbron, H., editors. 2008. *Standards and Thresholds for Impact Assessment.* Berlin: Springer Berlin Heidelberg. 493 p.

Schwartz, M.O. 2015. Modeling the hypothetical methane-leakage in a shale-gas project and the impact on groundwater quality. *Environmental Earth Sciences* 73(8): 4619–4632.

Scott, N.A., Tate, K.R., Giltrap, D.J., Tattersall Smith, C., Wilde, H.R., and New-some, P.J.F., and Davis, M.R. 2002. Monitoring land-use change effects on soil carbon in New Zealand: Quantifying baseline soil carbon stocks. *Environmental Pollution* 116(1): S167–S186.

Scott, V.E., and Oldemeyer, J.L. 1983. Cavity-nesting bird requirements and response to snag cutting in ponderosa pine. In: *Snag Habitat Management: Proceedings of the Symposium*. General Technical Report RM-99. Davis, J.W., Goodwin, G.A., and Ockentels, R.A., editors. Fort Collins: United States Department of Agriculture. 19–23 p.

Sebastián, C.R., and McClanahan, T.R. 2012. Using an ecosystem model to evaluate fisheries management options to mitigate climate change impacts in western Indian Ocean coral reefs. *Western Indian Ocean Journal of Marine Science* 11(1): 77–86.

Seber, G.A.F. 1982. *The Estimation of Animal Abundance and Related Parameters*. Second ed. New York: Palgrave Macmillan. 654 p.

Seidl, R., Rammer, W., Jäger, D., and Lexer, M.J. 2008. Impact of bark beetle (*Ips typographus* L.) disturbance on timber production and carbon sequestration in different management strategies under climate change. *Forest Ecology and Management* 256(3): 209–220.

Sellers, P., Hall, F., Asrar, G., Strebel, D., and Murphy, R. 1988. The first ISLSCP field experiment (FIFE). *Bulletin of the American Meteorological Society* 69(1): 22–27.

Sellers, P.J., Mintz, Y., Sud, Y.C., and Dalcher, A. 1986. A simple biosphere model (SiB) for use within general circulation models. *Journal of the Atmospheric Sciences* 43(6): 505–531.

Selong, J.H., and Helfrich, L.A. 1998. Impacts of trout culture effluent on water quality and biotic communities in Virginia headwater streams. *The Progressive Fish-Culturist* 60(4): 247–262.

Seymour, R.S., White, A.S., and deMaynadier, P.G. 2002. Natural disturbance regimes in northeastern North America – evaluating silvicultural systems using natural scales and frequencies. *Forest Ecology and Management* 155(1–3): 357–367.

Sharma, R. 2001. Indian deep-sea environment experiment (INDEX): An appraisal. *Deep Sea Res Part 2: Top Stud Oceanogr* 48(16): 3295–3307.

Sharma, R.K., Buffington, J.D., and McFadden, J.T. 1976. *The Biological Significance of Environmental Impacts*. Conference proceedings. NR-CONF-002. Washington, DC: US Nuclear Regulatory Commission. 327 p.

Shepherd, A., and Bowler, C. 1997. Beyond the requirements: Improving public participation in EIA. *Journal of Environmental Planning and Management* 40(6): 725–738.

Shifley, S.R., Thompson III, F.R., Dijak, W.D., Larson, M.A., and Millspaugh, J.J. 2006. Simulated effects of forest management alternatives on landscape structure and habitat suitability in the Midwestern United States. *Forest Ecology and Management* 229(1–3): 361–377.

Shugart, H.H., Goldstein, R.A., O'Neill, R.V., and Mankin, J.B. 1974. TEEM: A terrestrial ecosystem energy model for forests. *Oecologica Plantarium* 9: 251–284.

Shugart, H.H., Hopkins, M.S., Burgess, I.P., and Mortlock, A.T. 1980. Development of a succession model for subtropical rain forest and its application to assess the effects of timber harvest at Wiangaree State Forest, New South Wales. *Journal of Environmental Management* 11: 243–265.

Shugart, H.H., and West, D.C. 1977. Development of an Appalachian deciduous forest succession model and its application to assessment of the impact of the chestnut blight. *Journal of Environmental Management* 5: 161–179.

Shure, D.J. 1971. Insecticide effects on early succession in an old field ecosystem. *Ecology* 52(2): 271–279.

Shuter, B.J., Wismer, D.A., Regier, H.A., and Matuszek, J.E. 1985. An application of ecological modelling: Impact of thermal effluent on a smallmouth bass population. *Transactions of the American Fisheries Society* 114(5): 631–651.

Silvert, W. 1997. Ecological impact classification with fuzzy sets. *Ecological Modeling* 96(1): 1–10.

Sinclair, A.J., and Diduck, A. 1995. Public education: An undervalued component of the environmental assessment public involvement process. *Environmental Impact Assessment Review* 15(3): 219–240.

Sippe, R. 1999. Criteria and standards for assessing significant impact. In: *Handbook of Environmental Impact Assessment*. Volume 1. Petts, J., editor. Oxford: Blackwell Science. 74–92 p.

Sklar, F.H., Costanza, R., and Day, J.W. 1985. Dynamic spatial simulation modeling of coastal wetland habitat succession. *Ecological Modeling* 29(1): 261–281.

Snell, T., and Cowell, R. 2006. Scoping in environmental impact assessment: Balancing precaution and efficiency? *Environmental Impact Assessment Review* 26(4): 359–376.

Soer, G.J.R. 1980. Estimation of regional evapotranspiration and soil moisture conditions using remotely sensed crop surface temperatures. *Remote Sensing of Environment* 9(1): 27–45.

Soetaert, K., and Herman, P. 2009. *A Practical Guide to Ecological Modeling: Using R as a Simulation Platform*. Berlin: Springer. 372 p.

Solomon, A.M. 1986. Transient response of forests to CO_2-induced climate change: Simulation modeling experiments in eastern North America. *Oecologia* 68(4): 567–579.

Solomon, S.I., and Gupta, S.K. 1977. Distributed numerical model for estimating runoff and sediment discharge of ungaged rivers 2. model development. *Water Resources Research* 13(3): 619–629.

Sonntag, N.C., Everitt, R.R., Rattie, L.P., Colnett, C.L., Wolf, C.P., Truett, J.C., Dorcey, A.H.J., and Holling, C.S. 1987. *Cumulative Effects Assessment: A Context for Research and Development*. Hull: Canadian Environmental Assessment Research Council. 91 p.

Sorensen, J.H., Soderstrom, J., and Carnes, S.A. 1984. Sweet for the sour: Incentives in environmental mediation. *Environmental Management* 8(4): 287–294.

Sorensen, T., McLoughlin, P.D., Hervieux, D., Dzus, E., Nolan, J., Wynes, B., and Boutin, S. 2008. Determining sustainable levels of cumulative effects for boreal caribou. *Journal of Wildlife Management* 72(4): 900–905.

Spaulding, M.L., Saila, S.B., Lorda, E., Walker, H., Anderson, E., and Swanson, J.C. 1983. Oil-spill fishery impact assessment model: Application to selected Georges Bank fish species. *Estuarine, Coastal and Shelf Science* 16(5): 511–541.

Squires, A.J., Westbrook, C.J., and Dubé, M.G. 2010. An approach for assessing cumulative effects in a model river, the Athabasca River basin. *Integrated Environmental Assessment Management* 6(1): 119–134.

Stakhiv, E.Z. 1988. An evaluation paradigm for cumulative impact analysis. *Environmental Management* 12(5): 725–748.

Steenberg, J.W.N., Duinker, P.N., and Bush, P.G. 2011. Exploring adaptation to climate change in the forests of central Nova Scotia, Canada. *Forest Ecology and Management* 262(12): 2316–2327.

Steinemann, A. 2001. Improving alternatives for environmental impact assessment. *Environmental Impact Assessment Review* 21(1): 3–21.

Stevenson, M.G. 1996. Indigenous knowledge in environmental assessment. *Arctic* 49(3): 278–291.

Stewart, J.M.P., and Sinclair, J.A. 2007. Meaningful public participation in environmental assessment: Perspectives from Canadian participants, proponents, and government. *Journal of Environmental Assessment Policy and Management* 9(2): 161–183.

Stewart-Oaten, A., Murdoch, W.W., and Parker, K.R. 1986. Environmental impact assessment: "Pseudoreplication" in time? *Ecology* 67(4): 929–940.

Strikwerda, T.E., Fuller, M.R., Seegar, W.S., Howey, P.W., and Black, H.D. 1986. Bird-borne satellite transmitter and location program. *Johns Hopkins APL Technical Digest* 7(2): 203–208.

Sullivan, N.H., Bolstad, P.V., and Vose, J.M. 1996. Estimates of net photosynthetic parameters for twelve tree species in mature forests of the southern Appalachians. *Tree Physiology* 16(4): 397–406.

Suter, G.W. 1982. Terrestrial perturbation experiments for environmental assessment. *Environmental Management* 6(1): 43–54.

Swank, W.T., and Douglass, J.E. 1975. Nutrient flux in undisturbed and manipulated forest ecosystems in the southern Appalachian Mountains. In: *Proceedings of the Tokyo Symposium on the Hydrological Characteristics of River Basins and the Effects of these Characteristics on Better Water Management*. Tokyo: International Association of Hydrological Sciences. 445–456 p.

Tate, K.R., Scott, N.A., Saggar, S., Giltrap, D.J., Baisden, W.T., Newsome, P.F., Trotter, C.M., and Wilde, R.H. 2003. Land-use change alters New Zealand's terrestrial carbon budget: Uncertainties associated with estimates of soil carbon change between 1990–2000. *Tellus B* 55(2): 364–377.

Tennøy, A., Kværner, J., and Gjerstad, K.I. 2006. Uncertainty in environmental impact assessment predictions: The need for better communication and more transparency. *Impact Assessment Project Appraisal* 24(1): 45–56.

Thiel, H., Schriever, G., Ahnert, A., Bluhm, H., Borowski, C., and Vopel, K. 2001. The large-scale environmental impact experiment DISCOL – reflection and foresight. *Deep Sea Res Part 2: Top Stud Oceanogr* 48(17–18): 3869–3882.

Thompson, M.A. 1990. Determining impact significance in EIA: A review of 24 methodologies. *Journal of Environmental Management* 30(3): 235–250.

Toro, J., Requena, I., Duarte, O., and Zamorano, M. 2013. A qualitative method proposal to improve environmental impact assessment. *Environmental Impact Assessment Review* 43: 9–20.

Trent, T.T., and Rongstad, O.J. 1974. Home range and survival of cottontail rabbits in southwestern Wisconsin. *Journal of Wildlife Management* 38(3): 459–472.

Treweek, J. 1999. *Ecological Impact Assessment*. Oxford: Blackwell Science. 351 p.

Tucker, C.J. 1979. Red and photographic infrared linear combinations for monitoring vegetation. *Remote Sensing of Environment* 8(2): 127–150.

Usher, P.J. 2000. Traditional ecological knowledge in environmental assessment and management. *Arctic* 53(2): 183–193.

Valve, H. 1999. Frame conflicts and the formulation of alternatives: Environmental assessment of an infrastructure plan. *Environmental Impact Assessment Review* 19(2): 125–142.

Van Ballenberghe, V., and Peek, J.M. 1971. Radiotelemetry studies of moose in northeastern Minnesota. *Journal of Wildlife Management* 35(1): 63–71.

van Breda, L.M., and Dijkema, G.P.J. 1998. EIA's contribution to environmental decision-making on large chemical plants. *Environmental Impact Assessment Review* 18(4): 391–410.

van Breemen, N., Jenkins, A., Wright, R.F., Beerling, D.J., Arp, W.J., Berendse, F., Beier, C., Collins, R., van Dam, D., Rasmussen, L., et al. 1998. Impacts of elevated carbon dioxide and temperature on a boreal forest ecosystem (CLIMEX Project). *Ecosystems* 1(4): 345–351.

van der Valk, A.G., Squires, L., and Welling, C.H. 1994. Assessing the impacts of an increase in water level on wetland vegetation. *Ecological Applications* 4(3): 525–534.

Van Winkle, W., Jager, H.I., Railsback, S.F., Holcomb, B.D., Studley, T.K., and Baldrige, J.E. 1998. Individual-based model of sympatric populations of brown and rainbow trout for instream flow assessment: Model description and calibration. *Ecological Modeling* 110(2): 175–207.

Walker, C.R. 1976. Polychlorinated biphenyl compounds (PCBs) and fishery resources. *Fisheries* 1(4): 19–25.

Walker, S.L., Hedley, K., and Porter, E. 2002. Pulp and paper environmental effects monitoring in Canada: an overview. *Water Quality Research Journal of Canada* 37(1): 7–19.

Walters, C. 1974. An interdisciplinary approach to development of watershed simulation models. *Technological Forecasting and Social Change* 6: 299–323.

Walters, C.J. 1986. *Adaptive Management of Renewable Resources*. New York: Palgrave Macmillan. 374 p.

Walters, C.J. 1993. Dynamic models and large scale field experiments in environmental impact assessment and management. *Australian Journal of Ecology* 18(1): 53–61.

Walters, C.J., and Efford, I.E. 1972. Systems analysis in the Marion Lake IBP Project. *Oecologia* 11(1): 33–44.

Walters, C.J., Hilborn, R., and Peterman, R. 1975. Computer simulation of barren-ground caribou dynamics. *Ecological Modeling* 1(4): 303–315.

Walters, C.J., and Holling, C.S. 1990. Large-scale management experiments and learning by doing. *Ecology* 71(6): 2060–2068.

132   *Science in the EIA process*

Walters, C., Pauly, D., and Christensen, V. 1999. Ecospace: Prediction of mesoscale spatial patterns in trophic relationships of exploited ecosystems, with emphasis on the impacts of marine protected areas. *Ecosystems* 2(6): 539–554.

Walters, C.J., Steer, G., and Spangler, G. 1980. Responses of lake trout (*Salvelinus namaycush*) to harvesting, stocking, and lamprey reduction. *Canadian Journal of Fisheries and Aquatic Sciences* 37(11): 2133–2145.

Ward, D.V. 1978. *Biological Environmental Impact Studies: Theory and Methods.* New York: Academic Press. 157 p.

Wardekker, J.A., van der Sluijs, J.P., Janssen, P.H.M., Kloprogge, P., and Petersen, A.C. 2008. Uncertainty communication in environmental assessments: Views from the Dutch science-policy interface. *Environmental Science & Policy* 11(7): 627–641.

Warner, N.R., Christie, C.A., Jackson, R.B., and Vengosh, A. 2013. Impacts of shale gas wastewater disposal on water quality in Western Pennsylvania. *Environmental Science & Technology* 47(20): 11849–11857.

Webster, J.R., Blood, E.R., Gregory, S.V., Gurtz, M.E., Sparks, R.E., and Thurman, E.M. 1985. Long-term research in stream ecology. *Bulletin of the Ecological Society of America* 66(3): 346–353.

Weclaw, P., and Hudson, R.J. 2004. Simulation of conservation and management of woodland caribou. *Ecological Modeling* 177(1–2): 75–94.

Westman, W.E. 1985. *Ecology, Impact Assessment, and Environmental Planning.* New York: John Wiley & Sons. 532 p.

Weston, J. 2000. EIA, decision-making theory and screening and scoping in UK practice. *Journal of Environmental Planning and Management* 43(2): 185–203.

White, C.M., Thurow, T.L., and Sullivan, J.F. 1979. Effects of controlled disturbance on ferruginous hawks as may occur during geothermal energy development. In: *Transactions.* Volume 3. Davis: Geothermal Resources Council. 777–779 p.

Whittaker, R.H., Likens, G.E., Bormann, F.H., Easton, J.S., and Siccama, T.G. 1979. The Hubbard Brook ecosystem study: Forest nutrient cycling and element behavior. *Ecology* 60(1): 203–220.

Whittaker, R.H., Likens, G.E., Bormann, F.H., Easton, J.S., and Siccama, T.G. 1979. The Hubbard Brook ecosystem study: forest nutrient cycling and element behavior. *Ecology* 60(1): 203–220.

Whittington, J., St. Clair, C.C., and Mercer, G. 2005. Spatial responses of wolves to roads and trails in mountain valleys. *Ecological Applications* 15(2): 543–553.

Wich, S.A., Utami-Atmoko, S.S., Setia, T.M., Rijksen, H.D., Schürmann, C., van Hooff, JARAM, van Schaik, C.P. 2004. Life history of wild Sumatran orangutans (*Pongo abelii*). *Journal of Human Evolution* 47(6): 385–398.

Wichelman, A.F. 1976. Administrative agency implementation of the National Environmental Policy Act of 1969: A conceptual framework for explaining differential response. *Natural Resource Journal* 16: 263–300.

Wiens, J.A., Crawford, C.S., and Gosz, JR. 1985. Boundary dynamics: A conceptual framework for studying landscape ecosystems. *Oikos* 45(3): 421–427.

Wilbur, H.M. 1975. The evolutionary and mathematical demography of the turtle *Chrysemys picta*. *Ecology* 56(1): 64–77.

Wild-Allen, K., Herzfeld, M., Thompson, P.A., Rosebrock, U., Parslow, J., and Volkman, J.K. 2010. Applied coastal biogeochemical modeling to quantify the environmental impact of fish farm nutrients and inform managers. *Journal of Marine Systems* 81(1–2): 134–147.

Williams, G.L. 1981. An example of simulation models as decision tools in wildlife management. *Wildlife Society B* 9(2): 101–107.

Winter, T.C. 1978. Numerical simulation of steady state three-dimensional groundwater flow near lakes. *Water Resources Research* 14(2): 245–254.

Witmer, G.W., and deCalesta, D.S. 1985. Effect of forest roads on habitat use by Roosevelt elk. *Northwest Science* 59(2): 122–125.

Wood, C. 1995. *Environmental Impact Assessment: A Comparative Review*. Harlow: Longman Scientific & Technical. 337 p.

Wood, G. 2008. Thresholds and criteria for evaluating and communicating impact significance in environmental statements: 'See no evil, hear no evil, speak no evil'? *Environmental Impact Assessment Review* 28(1): 22–38.

Wood, G., Glasson, J., and Becker, J. 2006. EIA scoping in England and Wales: Practitioner approaches, perspectives and constraints. *Environmental Impact Assessment Review* 26(3): 221–241.

Wood, G., Rodriguez-Bachiller, A., and Becker, J. 2007. Fuzzy sets and simulated environmental change: Evaluating and communicating impact significance in environmental impact assessment. *Environment and Planning A* 39(4): 810–829.

Wright, R.F. 1998. Effect of increased carbon dioxide and temperature on runoff chemistry at a forested catchment in southern Norway (CLIMEX Project). *Ecosystems* 1(2): 216–225.

Wright, R.F., Lotse, E., and Semb, A. 1988. Reversibility of acidification shown by whole-catchment experiments. *Nature* 334(6184): 670–675.

Zarnoch, S.J., and Turner, B.J. 1974. A computer simulation model for the study of wolf-moose population dynamics. *Journal of Environmental Systems* 4(1): 39–51.

Zarnowitz, J.E., and Manuwal, D.A. 1985. The effects of forest management on cavity-nesting birds in northwestern Washington. *Journal of Wildlife Management* 49(1): 255–263.

Zavattaro, L., and Grignani, C. 2001. Deriving hydrological parameters for modeling water flow under field conditions. *Soil Science Society of America Journal* 65(3): 655–667.

8 Conclusions

Previous sections have attempted to address two broad aims: (i) to provide an overview of scientific developments associated with EIA since the early 1970s, as evidenced in the peer-reviewed literature, and (ii) to judge, on the basis of evidence found in the literature review, whether scientific theory and practice are at their vanguard in EIA-related applications. Here we present a summary of our findings and point to some promising avenues for future research.

Science has been an integral component of EIA since its inception in the 1970s. Though this review has uncovered several advancements surrounding EIA-related science since that time, it has also confirmed the continued relevance of first- and second-generation guidance materials (e.g., Holling 1978; Beanlands and Duinker 1983). Frameworks for handling scientific uncertainty in EIA (e.g., adaptive management, post-normal science) have remained influential in much of the literature reviewed, but appear to have had limited application in the context of formal EIA practice. Most importantly, perhaps, emerging ecological concepts and imperatives like biodiversity and climate change have begun to expand EIA's focus on VECs that are typically selected on the basis of traditional ecosystem elements (e.g., water, air, wildlife). At the same time, related concepts like resilience, thresholds, complexity, and landscape ecology appear to be serving EIA studies only in the background. Despite ample evidence of such conceptual and technical developments in the scientific peer-reviewed literature, the general consensus among scholars seems to be that science in EIA practice has not kept pace with these advancements.

Attempting to characterize and evaluate scientific advancements in EIA practice since the 1970s proved to be somewhat more difficult than our first objective, since most regulatory EIA filings do not find their way into the scholarly literature. We did, however, find that most academic reviews have continued to criticize the quality of science practiced in EIA since the 1980s (e.g., Fairweather 1994; Treweek 1995; Warnken and Buckley 1998; Greig and Duinker 2011). At the same time, many practitioners have reported feeling satisfied with the quality of science in EIA, but dissatisfied with the level of importance placed upon it by decision-makers (e.g., Morrison-Saunders and Bailey 2003; Morrison-Saunders and Sadler 2010). We propose that a more robust inquiry into the quality of science in EIA would rely on multiple lines of evidence, including that generated through practitioner surveys, workshops, and EIA document reviews.

In addition to highlighting the need for more empirical research around the question of scientific quality in EIA, the review has identified several promising avenues for future research in EIA-related science:

(i) delineation of basic cause-effect relationships linking anthropogenic stressors to VECs using both field experiments and monitoring programs (e.g., Johnson et al. 2005; Houle et al. 2010);

(ii) assembly of cause-effect knowledge into predictive simulation models that can be used to forecast environmental impacts (e.g., Weclaw and Hudson 2004; Kurz et al. 2008);

(iii) characterization of ecological thresholds associated with the condition of particular VECs (e.g., Richardson et al. 2007; Sorensen et al. 2008);

(iv) collaborative methods for engaging stakeholders throughout various scientific stages of the EIA process (e.g., Mulvihill 2003; Videira et al. 2010; Bond et al. 2015);

(v) integration of formal scientific knowledge with other forms of knowledge, such as traditional and Aboriginal ecological knowledge (e.g., Huntington et al. 2004a, 2004b; Gagnon and Berteaux 2009).

Contemporary debates surrounding science in EIA have attributed science-related challenges to several factors. One long-standing argument has been that poor science in EIA is the result of inadequate scoping (e.g., Ross et al. 2006; Morrison-Saunders et al. 2014). Others (e.g., Mulvihill 2003; Greig and Duinker 2014) have challenged this notion, pointing out that as long as EIA remains a restrictive, proponent-dominated endeavour, scientific contributions and decision outcomes will remain less than satisfactory. Meanwhile, broader debates have centred on finding an appropriate role for science in EIA. Whereas some (e.g., Cashmore 2004) have challenged an ongoing role for science in an increasingly politically driven EIA process, others (e.g., Greig and Duinker 2011) have defended it, emphasizing its importance in providing useful and defensible predictions of environmental impact.

Based on this literature review, we are convinced that science remains critical to EIA's central task of protecting VEC sustainability, but that scientific practice in EIA has not kept pace with developments in impact-related science broadly conceived. We also believe that any improvements to the scientific enterprise in EIA will rely on the adoption of more collaborative, creative, and participatory arrangements for designing and implementing EIA-related scientific studies.

References

Beanlands, G.E., and Duinker, P.N. 1983. *An Ecological Framework for Environmental Impact Assessment in Canada*. Halifax: Institute for Resource and Environmental Studies, Dalhousie University. 132 p.

Bond, A.J., Morrison-Saunders, A., Gunn, J.A.E., Pope, J., and Retief, F. 2015. Managing uncertainty, ambiguity and ignorance in impact assessment by embedding evolutionary resilience, participatory modeling and adaptive management. *Journal of Environmental Management* 151: 97–104.

Cashmore, M. 2004. The role of science in environmental impact assessment: Process and procedure versus purpose in the development of theory. *Environmental Impact Assessment Review* 24(4): 403–426.

Fairweather, P. 1994. Improving the use of science in environmental assessments. *Australian Journal of Zoology* 29(3–4): 217–223.

Gagnon, C.A., and Berteaux, D. 2009. Integrating traditional ecological knowledge and ecological science: A question of scale. *Ecology and Society* 14(2): 19.

Greig, L.A., and Duinker, P.N. 2011. A proposal for further strengthening science in environmental impact assessment in Canada. *Impact Assessment Project Appraisal* 29(2): 159–165.

Greig, L.A., and Duinker, P.N. 2014. Strengthening impact assessment: What problems do integration and focus fix? *Impact Assessment Project Appraisal* 32(1): 23–24.

Holling, C.S., editor. 1978. *Adaptive Environmental Assessment and Management*. Toronto: John Wiley & Sons. 377 p.

Houle, M., Fortin, D., Dussault, C., Courtois, R., and Ouellet, J. 2010. Cumulative effects of forestry on habitat use by gray wolf (*Canis lupus*) in the boreal forest. *Landscape Ecology* 25(3): 419–433.

Huntington, H., Callaghan, T., Fox, S., and Krupnik, I. 2004a. Matching traditional and scientific observations to detect environmental change: A discussion on arctic terrestrial ecosystems. *Ambio* 33(7): 18–23.

Huntington, H.P., Suydam, R.S., and Rosenberg, D.H. 2004b. Traditional knowledge and satellite tracking as complementary approaches to ecological understanding. *Environmental Conservation* 31(3): 177–180.

Johnson, C.J., Boyce, M.S., Case, R.L., Cluff, H.D., Gau, R.J., Gunn, A., and Mulders, R. 2005. Cumulative effects of human developments on Arctic wildlife. *Wildlife Monograpjs* 160: 1–36.

Kurz, W.A., Dymond, C.C., Stinson, G., Rampley, G.J., Neilson, E.T., Carroll, A.L., Ebata, T., and Safranyik, L. 2008. Mountain pine beetle and forest carbon feedback to climate change. *Nature* 452(7190): 987–990.

Morrison-Saunders, A., and Bailey, J. 2003. Practitioner perspectives on the role of science in environmental impact assessment. *Environmental Management* 31(6): 683–695.

Morrison-Saunders, A., Pope, J., Gunn, J.A.E., Bond, A., and Retief, F. 2014. Strengthening impact assessment: a call for integration and focus. *Impact Assessment Project Appraisal* 32(1): 2–8.

Morrison-Saunders, A., and Sadler, B. 2010. The art and science of impact assessment: Results of a survey of IAIA members. *Impact Assessment Project Appraisal* 28(1): 77–82.

Mulvihill, P.R. 2003. Expanding the scoping community. *Environmental Impact Assessment Review* 23(1): 39–49.

Richardson, C.J., King, R.S., Qian, S.S., Vaithiyanathan, P., Qualls, R.G., and Stow, C.A. 2007. Estimating ecological threholds in the Everglades. *Environmental Science & Technology* 41: 8084–8091.

Ross, W.A., Morrison-Saunders, A., Marshall, R., Sánchez, L.E., Weston, J., Au, E., Morgan, R.K., Fuggle, R., Sadler, B., Ross, W.A., et al. 2006. Improving quality. *Impact Assessment Project Appraisal* 24(1): 3–22.

Sorensen, T., McLoughlin, P.D., Hervieux, D., Dzus, E., Nolan, J., Wynes, B., and Boutin, S. 2008. Determining sustainable levels of cumulative effects for boreal caribou. *Journal of Wildlife Management* 72(4): 900–905.

Treweek, J. 1995. Ecological impact assessment. *Impact Assessment* 13(3): 289–315.

Videira, N., Antunes, P, Santos, R., and Lopes, R. 2010. A participatory modeling approach to support integrated sustainability assessment processes. *Systems Research and Behavioral Science* 27(4): 446–460.

Warnken, J., and Buckley, R. 1998. Scientific quality of tourism environmental impact assessment. *Journal of Applied Ecology* 35(1): 1–8.

Weclaw, P., and Hudson, R.J. 2004. Simulation of conservation and management of woodland caribou. *Ecological Modeling* 177(1–2): 75–94.

Index

Milton Keynes UK
Ingram Content Group UK Ltd.
UKHW022358061024
449327UK00031B/2571